Gamma Knife Brain Surgery

Progress in Neurological Surgery

Vol. 14

Series Editor   *L.D. Lunsford*, Pittsburgh, Pa.

KARGER   Basel · Freiburg · Paris · London · New York ·
New Delhi · Bangkok · Singapore · Tokyo · Sydney

Gamma Knife Brain Surgery

Volume Editors   *L.D. Lunsford*, Pittsburgh, Pa.
                  D. Kondziolka, Pittsburgh, Pa.
                  J.C. Flickinger, Pittsburgh, Pa.

75 figures and 34 tables, 1998

Basel · Freiburg · Paris · London · New York ·
New Delhi · Bangkok · Singapore · Tokyo · Sydney

••••••••••••••••••••••

L. Dade Lunsford, MD, FACS

University of Pittsburgh
Medical Center
Pittsburgh, Pa., USA

Library of Congress Cataloging-in-Publication Data
Gamma knife brain surgery / volume editors, L.D. Lunsford, D. Kondziolka, J.C. Flickinger.
(Progress in neurological surgery; vol. 14)
Includes bibliographical references and indexes.
1. Brain – Surgery. 2. Radiosurgery. 3. Brain – Diseases. I. Lunsford, L. Dade.
II. Kondziolka, D. (Douglas), 1961– . III. Flickinger, J.C. (John C.) IV. Series.
[DNLM: 1. Radiosurgery. 2. Brain Diseases – surgery. W1 PR673 v. 14 1998WL 368 G193 1998]
RD594.15.G35 1998 617.4′81059–dc21
ISBN 3–8055–6637–9 (hardcover : alk. paper)

Bibliographic Indices. This publication is listed in bibliographic services, including Current Contents® and Index Medicus.

© Copyright 1998 by S. Karger AG, P.O. Box, CH–4009 Basel (Switzerland)
Printed in Switzerland on acid-free paper by Reinhardt Druck, Basel
ISBN 3–8055–6637–9

Contents

V

....

Preface

The series, *Progress in Neurological Surgery,* has a long and valuable educational history. The first volume was published by Karger in 1966. The initial distinguished editorial group included Krayenbuhl, Maspes and Sweet. Despite the more recent explosion in neurosurgical publishing over the last 21 years, the value and role of this series will be further shown by the publication of the present volume. When the publishers at Karger came to me with the concept of taking over the editorship, I suggested that the lapse of almost 8 years required rapid introduction of a timely and topical volume. I had been planning for some time to produce a volume summarizing our experience in stereotactic radiosurgery at the University of Pittsburgh Medical Center. A combination of these two goals allowed me to rapidly develop a series of contributing authors who could be 'leveraged' to produce a timely group of papers summarizing our experience.

Other guest-edited volumes of *Progress in Neurological Surgery* are expected on a biennial basis. I hope that readers will take advantage of the rejuvenation of this important series. I would like to express my deep appreciation to the entire team of colleagues who participated as authors in the book and wish to provide special acknowledgement to Charlene Baker, Mary Ann Vincenzini, Peggy Schmitt, and Stephanie Lunsford for their enormous contribution to the preparation of this book. Without them it could not and would not be done.

L. Dade Lunsford, MD, FACS

Lunsford LD, Kondziolka D, Flickinger JC (eds): Gamma Knife Brain Surgery.
Prog Neurol Surg. Basel, Karger, 1998, vol 14, pp 1–4

....

Introduction

Our interest in stereotactic radiosurgery began in the late 1970s after recognition of the unique potential of combining surgical guiding devices with the newly developed CT imaging. This interest was further enhanced by the senior editor's experience in Sweden in 1979 and 1980 where image-guided neurosurgery, functional neurosurgery and indeed stereotactic radiosurgery with the Gamma knife was flourishing. Image-guided brain surgery started at the University of Pittsburgh in 1981. The first North American 201 Cobalt-60 source Gamma knife was installed in 1987. Ten years have now passed. During this time, there has been an explosion in the local, regional, national and international interest in stereotactic radiosurgery.

By May 1987, 2,344 patients had undergone Gamma knife radiosurgery during our first 10-year experience (fig. 1). Initially, our case preponderance reflected vascular malformations for which there was preliminary outcome data. As time has passed, we have seen an increasing shift toward skull base tumors, malignant tumors including brain metastasis and selected glial neoplasms. More recently, our center and others here increased functional neurosurgery applications, especially for trigeminal neuralgia and for carefully selected movement disorder patients. At our center in Pittsburgh, gamma knife radiosurgery represents between 14 and 17% of the 3,250 neurosurgical procedures done at our institution. Our experience has increased from initially 185 patients per year to 425 patients per year on an annual basis. During this interval, we have pursued clinical research, academic research and outcome studies. The results of many of our evaluations and investigations are enclosed in the chapters herein.

The increase in our experience has been paralleled by the national and indeed international experience with Gamma knife radiosurgery. In the United States, 17,189 patients had undergone Gamma knife radiosurgery by June

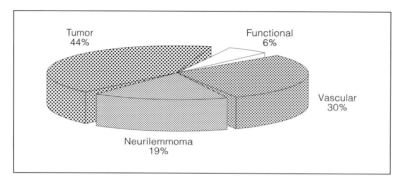

Fig. 1. The Pittsburgh Gamma knife experience from August 1987 through April 1997
(n = 2,344).

1997 (fig. 2). Approximately one-third of these tumors represented malignant
tumors, one-third represented benign tumors and almost 1,000 patients had
functional disorders including pain and Parkinson's disease. In 1992, 1,583
patients were treated at 11 sights. In 1997, 5,000 patients had been treated at
32 sites.

The average volume by year of operation is shown in figure 3. The pene-
trence of radiosurgery into the management of brain tumors varies. For
example, it is as low as 4% of patients with metastatic brain cancer who are
eligible for radiosurgery. In contrast, in 1997, approximately 25% of newly
diagnosed patients with acoustic neuromas (estimated to be approximately
2,200 patients per year in the United States) underwent Gamma knife radiosurgery. The growth of radiosurgery represents an exponential trend. While initially we projected exponential growth followed by a significant reduction
(in the usual fashion a decline in activity follows the recognition of adverse
complications). Instead, we have seen a continual escalation in both the safety
and the efficacy of radiosurgery.

Centers embarking on Gamma knife radiosurgery must demonstrate a
continuous multi-disciplinary commitment. Critical input must be provided
by specialists in neurological surgery, radiation oncology, and medical physics.
Commitment to high-resolution neurodiagnostic imaging is paramount. Patient selection issues are critical, especially to avoid a tendency to recommend
radiosurgery for large tumors or AVMs.

During the 10 years that we have worked with the Gamma knife, we have
had the distinct pleasure of working with 26 neurosurgeons who specifically
came to our center for training in stereotactic radiosurgery. They have come

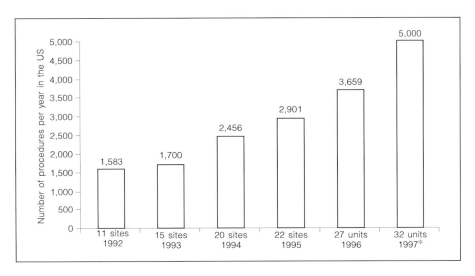

Fig. 2. The annual total of Gamma knife experience in the United States (1991–1997). *Projection based on annualized rate and number of clinical start-ups for the remainder of the year. (Survey of the Leksell Gamma Knife Society, June 1997.)

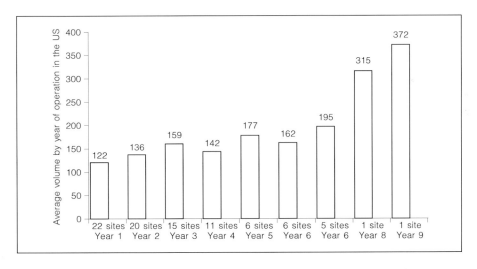

Fig. 3. The average volume of Gamma knife brain surgery by year of operation. (Survey of the Leksell Gamma Knife Society, June 1997.)

from North America, South America, Europe, Asia and Africa. Almost all are currently working with radiosurgical techniques in their own environment. We have prevailed upon several of them to demonstrate their expertise and write a chapter using the comprehensive radiosurgical database that we have maintained at the University of Pittsburgh. It is our hope that the reader will be able to use this data to assess the overall impact of stereotactic radiosurgery in the management of a wide variety of neurological disorders.

L. Dade Lunsford, MD, FACS
Douglas Kondziolka, MD, MSc, FRCS(C)
John C. Flickinger, MD

Lunsford LD, Kondziolka D, Flickinger JC (eds): Gamma Knife Brain Surgery.
Prog Neurol Surg. Basel, Karger, 1998, vol 14, pp 5–20

....

Gamma Knife Technology and Physics: Past, Present, and Future

Ann Maitz, John C. Flickinger, L. Dade Lunsford

Center for Image-Guided Neurosurgery, University of Pittsburgh Medical Center,
Pittsburgh, Pa., USA

The role of stereotactic radiosurgery has expanded continuously. The installation and use of both Gamma knife units and modified linear accelerators are becoming more prevalent. This report discusses the important physics and licensing issues regarding only the Leksell Gamma knife. The fundamental design of the Gamma knife has changed only minimally during an almost 30-year experience. Proper use of the device is dependent on facility design and knowledge about licensing, shielding, government regulations, isodose contours, beam profiles, calibration, quality assurance, dosimetry (isocenter-isocenter interactions, blocking, quality control), commissioning, and quality management. The differences between the roles of the models of Gamma knife (U and B) also must be understood. Continuous advances and modifications of hardware and software have characterized the evolving role of the Gamma knife.

Unit Design

Both the Model U and the Model B Gamma knife units utilize 201 separate ^{60}Co sources focused to a single point that is 40.3 cm source to focus distance. ^{60}Co decays with two photons (1.17 and 1.33 MeV, average 1.25). Each one of the cobalt sources (General Electric, Vallecitos, Calif., USA) is 1 mm in diameter. Twenty 1-mm long pellets of ^{60}Co are stacked into a capsule. That capsule is inserted into another steel capsule which is enclosed by a bushing. The bushing is then loaded into the central body of the unit. The sources in the U model are grouped in an almost hemispherical array (fig. 1).

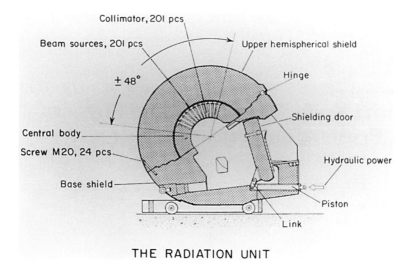

Collimator, 201 pcs

Beam sources, 201 pcs

Upper hemispherical shield

± 48°

Hinge

Shielding door

Central body

Screw M20, 24 pcs

Hydraulic power

Base shield

Piston

Link

THE RADIATION UNIT

Fig. 1. Cross section of the model U Gamma knife. The sources are located in a hemispherical array.

The central beam lies at a 55° angle to the horizontal plane while other beams are arranged in an arc of $\pm/48°$ about the x-axis and $\pm/80°$ about the z-axis. The B model sources are arranged in a circular array which is $\pm/36°$ about the x-axis (fig. 2). The sources are arranged in a hand-in-glove manner so that each source bushing assembly in the central body of the unit is aligned with its precollimator (65 mm of tungsten alloy), stationary collimator (92.5 mm of lead) and the final collimator (60 mm of tungsten alloy) on the helmet (fig. 3). The alignment of the sources with the final collimator helmet is ensured by the strict quality control exercised in the machining of the units. This alignment is verified during the commissioning of each unit. The exchangeable collimator helmets exist in four sizes (4, 8, 14 and 18 mm diameter). These diameters refer to the nominal diameter at the 50% isodose line. The measured diameters are shown in table 1.

Treatment Technique

The basic treatment technique is similar for both models. The Leksell frame is attached to the patient's head. The stereotactic frame center (x = 100, y = 100, z = 100) lies at the center of the frame in three-dimensional space. When applied to the patient, the frame is positioned so that the target is as close to the center of the frame in all three aspects as possible. Once the

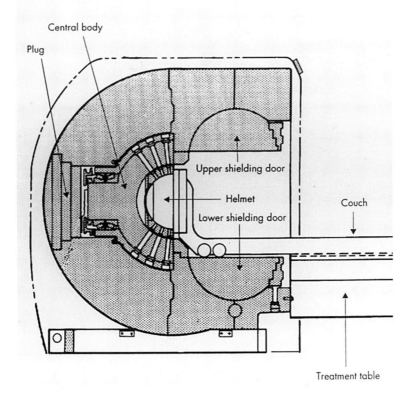

Central body

Plug

Upper shielding door

Helmet

Lower shielding door

Couch

Treatment table

Fig. 2. Cross section of the model B Gamma knife. The sources are arrayed in a circular 'donut' array.

patient's head is fixed within the frame, the patient is ready to undergo imaging followed by radiosurgery. During actual treatment, the surgeon first sets the Y (anterior-posterior) and Z (inferior-superior) coordinates of the stereotactic frame. The frame is then secured into the collimator helmet and support assembly by means of trunnions. The trunnions fix the X (right-left aspect) coordinate. After irradiation is initiated, the shielding door to the central body of the unit opens and the couch advances the patient so that the collimator helmet docks with the central body containing all 201 sources. As soon as the helmet docks with the central body, target irradiation commences. Once the treatment time has elapsed, the couch moves out and the shielding door closes.

A newly loaded Gamma knife has an activity of 6,000 Ci $\pm$/10% which produces a dose rate (in a 8-cm spherical radius phantom) of 300–400 cGy/min. Most centers find it necessary to reload after 8–10 years of use since ^{60}Co

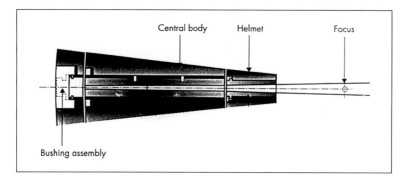

Fig. 3. The design of the bushing and helmet for a single photon beam of the Gamma knife.

Table 1. Dimensions of the 50% isodose (mm)

Collimator	U unit			B unit		
	x	y	z	x	y	z
18	23.7	21.0	25.0	25.0	28.0	20.6
14	18.4	16.4	19.5	20.5	22.0	16.0
8	10.6	9.7	11.7	12.0	12.5	9.5
4	6.0	6.0	6.2	6.0	6.5	5.0

decays with a half-life of 5.26 years. Loading or unloading of the U unit requires a temporary hot cell for shielding. The B unit has a specially designed loading device that makes the process less expensive and faster. In addition, it eliminates the need for special shielding during the loading process.

Licensing

In the United States, licensing and regulating the Gamma knife is the responsibility of either the Nuclear Regulatory Commission (NRC) or the individual state's Bureau of Radiation Health (in Agreement States). Agreement State status is conferred on those states whose regulations have been reviewed by the NRC and deemed adequate to protect the public and employees. On the whole, they do not differ significantly from the Code of Federal Regulations Title 10 (10 CFR) Parts 19, 20, 35 and 49 [2–5]. Prior to sale, the Gamma knife was approved by the Food and Drug Administration (FDA) for sale via the 501(k) premarket notification process.

For each Gamma knife site, a license report must be submitted. These aspects include designation of the authorized users, radiation safety training, development of operating procedures, and satisfactory emergency procedures. Safety checks must be performed on a daily, monthly, semiannual, and annual basis. Personnel radiation badge monitoring, loading and unloading procedures, quality management program, calibration procedures, leak test procedures, and shielding design must all be evaluated prior to license approval [7, 11].

The licensing of the first Gamma knife in the United States was a complex task [7]. The interactive information exchange at our site required almost 2 years. Eventually our constant communication with NRC reviewers and inspectors proved productive. Since transporting a loaded unit in the United States was prohibited, the first Gamma knife approved for clinical use was loaded on site [8]. This was the first time a Gamma knife actually was loaded on site. Many issues had to be resolved prior to obtaining permission to load the ^{60}Co sources into the unit. Because of the emergence of a specially constructed loading machine to facilitate reloading of the B unit (and augmented by established procedures that were performed many times worldwide without incident), the licensing procedure became relatively quick and painless.

Regulations

The regulations set forth by the NRC define the criteria for a misadministration as well as all required specific safety precautions [4]. Monthly spot checks include interlock tests, output checks, and a comparison of measured output with anticipated output. Beam status indicator checks, timer checks regarding termination of exposure, linearity, reproducibility, accuracy and error are tested as well. Emergency off buttons, adequacy of audiovisual communication with the patient, radiation monitors, and appropriate regulation notifications are also assessed at this time. Annual calibrations are also required. Such calibrations include tests of radiation isocenter/mechanical isocenter coincidence, beam output, dose profiles and emergency training.

Section 10 of the Code of Federal Regulations lists the qualifications of the authorized users and physicists who are named on the state license. Some neurosurgeons who operate units are not listed on the license. Those physicians or physicists who wish to be named on the license must be approved both by their institution's radiation safety committee and the medical review board of the NRC or state. Issues of public safety, such as leakage and scatter radiation, are also specified in radiation regulations. Health care facilities have radiation safety programs monitored by radiation safety committees.

Shielding

The shielding necessary to protect health care workers near the Gamma knife unit depends on the anticipated caseload [9, 11]. The output and scatter patterns from the unit, position and distance of the walls from the central body of the unit are taken into consideration. The radiation exposure rates adjacent to structures that reside on the exterior side of the shielding walls must be assessed as to the extent of radiation. The normal scatter patterns that occur for ^{60}Co teletherapy units and linear accelerators do not predict the levels that occur for the U-type Gamma knife (fig. 4, 5) or for the B-type Gamma knife (fig. 6, 7).

Quality Assurance

Calibrations, monthly spot checks and daily checks have been required as parts of the Gamma knife maintenance since 1987 [1, 9]. These checks give the physicist an opportunity to become familiar with the unit and detect small problems before they expand into larger ones. The checks are intended to maintain the safety of the general population, patients and employees. Except for the change from the hydraulically driven U unit to the electrically driven B unit, the items covered by the required checks have varied little through the years. However, monitoring techniques have become more accurate and sophisticated. This is largely due to the development of new detectors that are capable of measuring very small radiation fields with great accuracy. The isodose contour determination methods have progressed from only basic radiation therapy film and TLD work to GAF Chromic foils [10], laser readers, different types of film, Mosfet detectors and bang polymer gels.

Both the U-type and B-type Gamma knives have advantages, primarily related to different beam profiles and isodose curves (fig. 8, 9). The dose distributions in the axial and coronal planes of the B unit have significantly changed (at the lower isodoses) without affecting the 50% dimension appreciably. This is an advantage of the B unit especially when planning pituitary tumors. To create a similar plan with the U unit, a complex beam-blocking pattern would be necessary. The advantage of the U unit can be seen in the sagittal view of figure 9. The inclination of the unit can be used to correspond to the shape of certain lesions. This can also be accomplished by adjusting the gamma angle on the B unit, a technique that is relatively simple. The accuracy of the Gamma knife, both B-type and U-type, is evident in the mechanical isocenter/radiation isocenter coincidence test (fig. 10). This has been consistent over time and across both models.

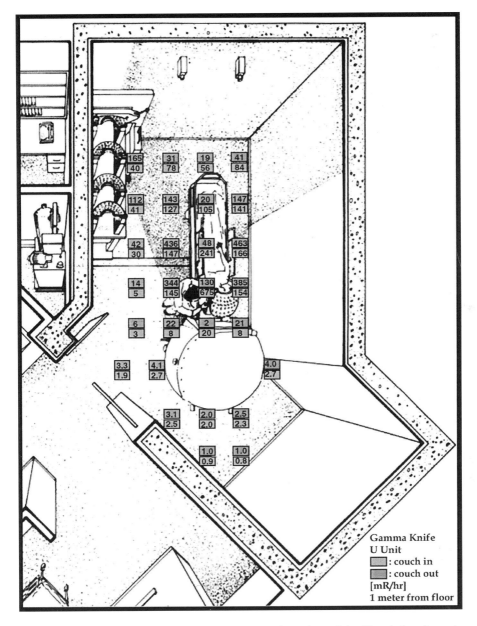

Fig. 4. Radiation exposure rates (MR/h) at various sites of the U unit 1 m from the floor.

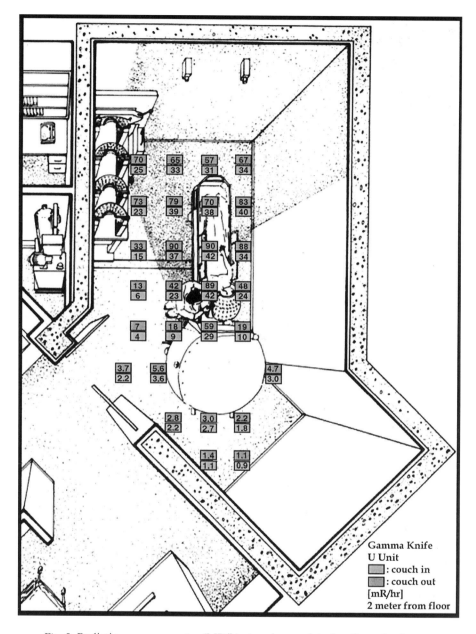

Fig. 5. Radiation exposure rates (MR/h) at various points 2 m from the floor.

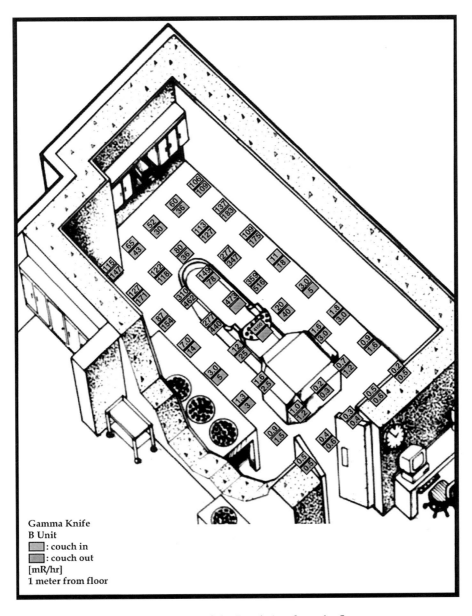

Fig. 6. Radiation exposure rates of the B unit 1 m from the floor.

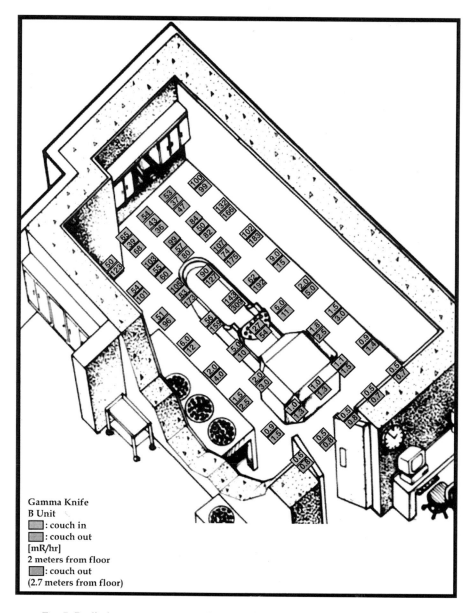

Fig. 7. Radiation exposure rates of the B unit 2.7 m from the floor.

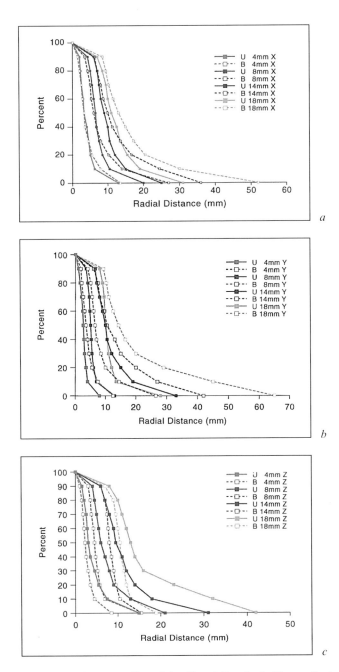

Fig. 8. Beam profiles of the U and B units in X coordinate (*a*), Y coordinate (*b*) and Z coordinate (*c*).

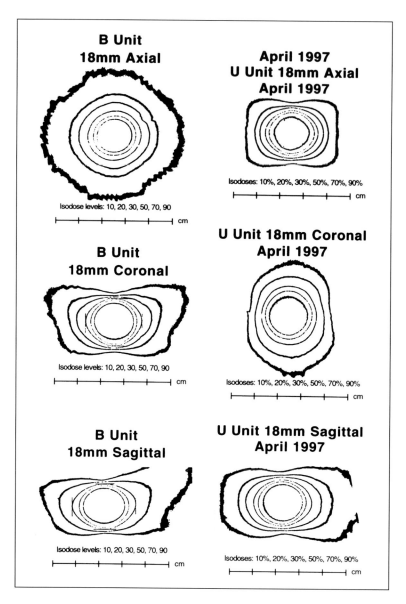

Fig. 9. Profiles of isodoses of the B unit (left) and U unit (right) in axial, coronal and sagittal planes.

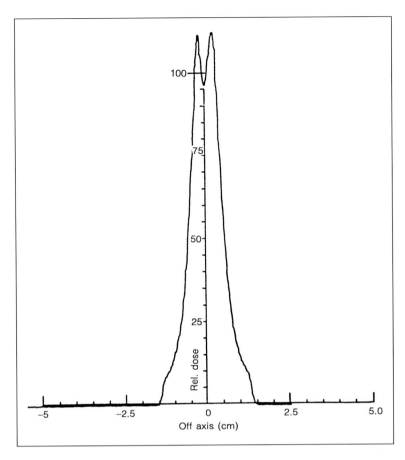

Fig. 10. The densitometer plot showing the coincidence of the radiation and mechanical isocenters of the Gamma knife.

Treatment Planning

The dose-planning aspect of Gamma knife stereotactic radiosurgery has changed dramatically through the years. The program first supplied with the U-type Gamma knife was named Kula® (AB Elekta, Stockholm, Sweden). It produced dose plots of the integration for all the planned isocenters. The plots were overlaid onto the hard copy x-ray films in order to verify the plan in comparison with the lesion. The dose-planning process thus far has been the development of Leksell Gamma Plan®. This program integrates the images with the isodose curves on screen and has reduced the time needed to calculate. Gamma Plan® reconstructs coronal and sagittal images from an axial acquisi-

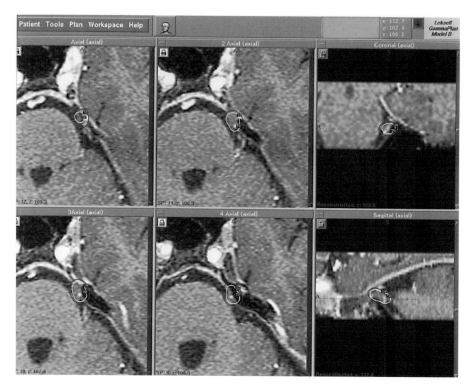

Patient Tools Plan Workspace Help

x: 112.7
y: 102.6
z: 108.2

*Leksell
GammaPlan
Model B*

Axial (axial)   2 Axial (axial)   Coronal (axial)

3Axial (axial)   4 Axial (axial)   Sagittal (axial)

Fig. 11. Axial and reformatted coronal and sagittal images of a two shot (4 × 2 mm). Plan for treatment of left trigeminal neuralgia.

tion (fig. 11). The patient is only subjected to one brief MRI acquisition sequence. The accuracy of the axial acquisition is assessed by quality assurance measurements which are made on the images. These assessments are made both at the imaging console and after import of the images to the treatment planning software. If the images are not within the acceptable range of error, the patient is re-imaged.

Understanding isocenter-isocenter interactions for the novice planner is relatively effortless using Version 4 software. Experimentation with isocenter movements requires a minimal amount of time [13]. The automatic blocking pattern feature in Leksell Gamma Plan® allows the planning novice to create a plan that accommodates critical areas of the brain. The accomplished planner can achieve exquisite plans quickly (conformal plans are crucial).

Software quality control is reasonably easy. Daily checks include measurements on the acquired images as well as an output check on a 'dummy' case. The monthly check measures relative helmet factors and the output check confirms the measured output. An inspection related to the number of occlusive blocks and output is also performed to ensure the output reflects the number of beams occluded. Upgrades to new versions of software are relatively simple to perform. There are several tests in addition to the monthly software checks that should be done when upgrading the software. They involve prone versus supine treatments, skull radii calculations, a variety of blocking outputs, correct calculation of volumes, correct use of attenuation coefficient, and the ability to receive images and scale them properly.

Quality Management

There is one additional aspect that is required by licensing authorities. A quality management program is mandated and consists of several features. Prior to the administration of the treatment, there must be a written directive. It contains the coordinate of each target along with the blocking pattern and collimator size, the exposure time, the target dose, and the total dose. The neurosurgeon, radiation oncologist, and medical physicist review, sign and date each plan [12]. A procedure must be established so that both the computer-generated dose calculations and the patient identification are verified. Before each new isocenter is treated, the coordinates must be at least double checked. A triple check significantly reduces the margin of error [6]. The quality management program also orders that acceptance test be performed on all treatment-planning computers prior to the first use of the system based on the specific applications and needs of the users. This is performed by a qualified teletherapy physicist.

Future Development

There continues to be improvements in the technique of Gamma knife radiosurgery. The next step in software development is what is commonly referred to as 'inverse treatment planning'. This would allow even a novice treatment planner to specify some of the critical parameters and have the computer design the plan based on the constraints given by the user. Such plans will require final modifications by the planning team.

The subsequent phase in the evolution of the Gamma knife hardware would be to have a device to automatically change the position of the patient to the set of coordinates of the next treatment isocenter. This would avoid the potential for human error in setting the coordinates, except of course for the initial treatment settings. In addition, multiple isocenter treatments will be facilitated and performed more quickly. With the advent of an automatic patient-positioning system, there will arise a need for physics verification. This will ensure that the prescription intended for a particular patient was indeed delivered as it was prescribed by the physician. A method for determining the dose delivered, should the system fail, must emerge in order to comply with the rules in the United States.

Although it is still young, this field has advanced with tremendous velocity. Advances reflect the teamwork approach of neurosurgeons, radiation oncologists, and physicists. As the field grows, the team concept must remain in order to optimize the use of future innovations.

References

1   American Association of Physicists in Medicine: AAPM Rep No 54: Stereotactic Radiosurgery, June 1995.
2   Code of Federal Regulations Title 10: Part 19.
3   Code of Federal Regulations Title 10: Part 20.
4   Code of Federal Regulations Title 10: Part 35.
5   Code of Federal Regulations Title 10: Part 49.
6   Flickinger JC, Lunsford LD, Kondziolka DS, Maitz AH: Potential human error in setting stereotactic coordinate for radiosurgery: Implications for quality assurance. Int J Radiat Oncol Biol Phys 1993; 27:397–401.
7   Guide for the Preparation of Applications for Licenses for Medical Teletherapy Programs. Draft Guide 1985.
8   Maitz AH, Lunsford LD, Wu A, Lindner G, Flickinger JC: Shielding requirements, on-site leading and acceptance testing of the Leksell Gamma Knife: The Pittsburgh Experience. Int J Radiat Oncol Biol Phys 1990;18:469–476.
9   Maitz AH, Wu A, Lunsford LD, Flickinger JC, Kondziolka DS, Bloomer WD: Quality assurance for Gamma Knife stereotactic radiosurgery. Int J Radiat Oncol Biol Phys 1995;32:1465–1471.
10 McLaughlin WL, Soares CG, Sayeg JA, McCullough EC, Kline RW, Wu A, Maitz AH: The use of a radiochromic detector for the determination of stereotactic radiosurgery dose characteristics. Med Phys 1994;21:379–388.
11 National Council on Radiation Protection and Measurements: NCRP Rep No 49: Structural Shielding Design and Evaluation for Medical Use of X-Rays and Gamma Rays of Energies up to 10 MeV.
12 Regulatory Guide 8.33: Quality Management Program, Oct 1991.
13 Wu A, Lindner G, Maitz AH, Kalend AM, Lunsford LD, Flickinger JC, Bloomer WD: Physics of Gamma Knife approach on convergent beams in stereotactic radiosurgery. Int J Radiat Oncol Biol Phys 1990;18:941–949.

Ann Maitz, MS, Medical Physics, Center for Image-Guided Neurosurgery,
University of Pittsburgh Medical Center,
Suite B-400, 200 Lothrop Street, Pittsburgh, PA 15213 (USA)

Lunsford LD, Kondziolka D, Flickinger JC (eds): Gamma Knife Brain Surgery.
Prog Neurol Surg. Basel, Karger, 1998, vol 14, pp 21–38

....

Radiobiologic Considerations in Gamma Knife Radiosurgery

Douglas Kondziolka [a,b], *L. Dade Lunsford* [a,b], *Ann Maitz* [b],
John C. Flickinger [a,b]

Departments of [a] Neurological Surgery and [b] Radiation Oncology,
University of Pittsburgh and the Center for Image-Guided Neurosurgery,
Presbyterian University Hospital, Pittsburgh, Pa., USA

Radiosurgery is the precise and complete destruction of a chosen target containing healthy and/or pathological cells, without significant concomitant or late radiation damage to adjacent tissues [28]. The initial radiosurgery concept of Leksell was for the treatment of functional neurologic disorders. Today, radiosurgery is used for a myriad of indications both of the cranium and spinal column. The current radiosurgery concept is that damage to normal tissue within the target volume is not a problem but a desired effect. In radiosurgery the physician does not attempt to spare some tissues and treat others, but to achieve a total destructive effect within the targeted volume.

Van der Kogel [45] wrote that effects of radiosurgery with single doses of radiation, in radiobiological concepts, were no different from the effects of fractionated irradiations. The difference between radiosurgery and radiotherapy generally is the size of the treatment volume. While volume is important, it is the surgeon's ability to deliver precise and accurate radiation to a defined target during one procedure that provides a powerful radiobiologic effect. This effect often is not identified after standard dose fractionated radiotherapy to target and surrounding brain. We believe that accurate targeting opens the door for the powerful radiobiologic effect of radiosurgery. Since fractionated radiation therapy treats a relatively large tissue volume that incorporates both lesion and normal parenchyma, the physician must exploit some therapeutic ratio that injures tumor cells but maintains normal brain integrity. Consequently, the delivered dose is often relatively low. Many tumors are considered somewhat radioresistant (meningiomas, schwannomas, melanomas, sarcomas) to these low doses. Precise and accurate radiation delivery in radiosurgery means that lesions which do not contain normal tissue can be irradiated at a high dose provided that only a fraction of the dose is received by the surrounding brain [27]. A steep fall-off in radiation delivery is required. These powerful doses transcend the modern concept of 'radiation resistance'. Meningiomas, schwannomas, melanomas and sarcomas respond favorably and consistently to radiosurgery [43].

Hall and Brenner [13] agreed with the use of radiosurgery for benign brain lesions such as arteriovenous malformations (AVM) and benign tumors. These authors questioned its role for the management of malignant tumors. They derived radiobiologic data to support their contention that the treatment of malignant tumors with a single radiation fraction would result in a suboptimal therapeutic ratio between tumor control and late effects. An improved ratio would be expected from fractionation [13]. Their argument was based on the concept that hypoxic cells could reestablish their oxygenated state and become more sensitive to irradiation during the course of fractionated radiotherapy. Since AVMs and benign tumors are late-responding tissues, nothing was felt to be gained by fractionation. The performance of stereotactic fractionated radiotherapy for acoustic tumors may lessen the radiobiological effect on cranial nerves [27] (perhaps to overcome dose planning and delivery issues posed by some radiosurgery systems). Fractionation is unlikely to increase the therapeutic schwannoma response.

We advocate radiosurgery as a boost to fractionated radiotherapy in the management of malignant gliomas and not as the sole radiation approach. Fractionated radiation may not improve the results of treatment of metastatic tumors. These benefits may be dose-related. Since whole-brain irradiation is limited by brain tolerance, a powerful effect on the small tumor and surrounding brain may not be possible without a high rate of tissue injury. When radiosurgery is used as the sole treatment for a solitary metastasis, the problems of whole-brain tolerance are eliminated and the physician can focus on the delivery of a high dose to the tumor itself. Such an approach provides excellent local control (approximately 90% for most tumors) with concomitant longer survival [9].

Larson et al. [27] placed the different targets for radiosurgery use into four categories. Category 1 included a late-responding target embedded within late-responding normal tissue (e.g. AVM). Radiosurgery was believed an appropriate biologic strategy with no advantage to fractionation. In category 2, a late-responding target was surrounded by late-responding normal tissue (i.e. meningioma or other tumor that does not invade normal brain parenchyma). Again this category was believed to be appropriate for radiosurgery (fig. 1). Category 3 included early-responding targets embedded within late-

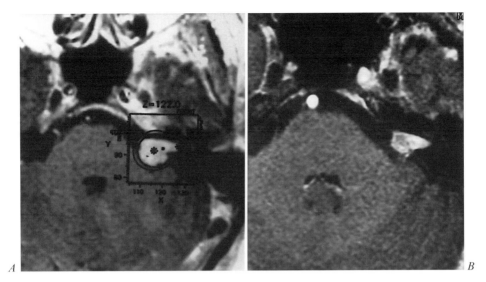

Fig. 1. Axial MR images at radiosurgery (*A*) and 4 years later (*B*) in a man with a small left acoustic tumor. After radiosurgery the tumor shows decreased contrast enhancement and has regressed in volume.

responding normal tissue (i.e. astrocytoma). In this tumor, both normal glial cells and neoplastic cells exist within the target volume. One might anticipate a poor therapeutic ratio with radiosurgery although our early reports on pilocytic astrocytoma have shown a favorable response with radiosurgery. For the most part, such tumors are of small volume and mainly in children where both the physician and family would like to avoid large-field irradiation in the developing brain. Category 4 included an early-responding tissue surrounded by a late-responding normal tissue (i.e. brain metastases). In this lesion, the target volume contains mainly malignant cells. Radiosurgery would be expected to kill oxygenated cells but might do less damage to hypoxic cells. As noted above, clinical reports do not support this concept since tumor control rates are high and morbidity rates low. Larson et al. [27] state that most fractionated regimens have not provided the radiobiologic effect that is administered during radiosurgery.

The Effect of Dose Rate

Reducing the dose rate of irradiation has an effect similar to fractionation. Over time, cells can repair sublethal radiation-induced damage [15]. The rate of repair (expressed as the half-time for disappearance of repairable injury) is approximately 1.5 h in the central nervous system [45]. The most obvious effect of dose rate is the decay in cobalt activity found with the Gamma knife. Since the half-life of cobalt is approximately 5 years, significant dose-rate effects are observed over time. Although we found no clinical differences during the first 9 years of use of the University of Pittsburgh Gamma knife prior to reloading, we did notice that cranial nerve morbidity rate during the first years were higher than expected, an observation that may have been related to a high dose-rate effect. Larson [26] hypothesized that the concept of microfractionation might be important in reducing morbidity simply by delaying treatment times between radiosurgery isocenters at different stereotactic coordinates (within the same overall procedure). A recent study by Shaw et al. [38] showed poorer tumor control rates with linear accelerator-based radiosurgery than with Gamma knife radiosurgery. One explanation for this result might be that the much lower dose rate of linear accelerator irradiation led to a reduced tumor biologic effect. Despite the potential importance of dose rate, no study has shown conclusively the effect of this factor. An analysis of postradiosurgery imaging changes in 307 AVM patients managed in Pittsburgh from 1987 to 1993 did not find any correlation with the dose rate at the periphery of the AVM nidus targeted [11]. The lack of correlation in this study indicates only that the dose-rate effect appears to be small within the range seen for the clinical use of the Gamma knife.

Experimental Comparisons between Radiosurgery and Fractionated Radiotherapy

Most clinicians do not have a good understanding of the magnitude of radiosurgical effects, especially when described in terms of conventional radiation therapy doses. Although it is difficult to select a dose for each group that is radiobiologically equivalent, use of information provided by Larson et al. [27] would be one starting point. In their report, for early-responding tissue such as a malignant neoplasm ($\alpha/\beta = 10$), a radiosurgery dose of 20 Gy was hypothesized to equal a fractionated dose of 50 Gy. Calculated α/β beta ratios for different malignant glioma cell lines showed a mean value of 10.4 [14, 30, 41, 48]. Thus, we realized that an α/β ratio of 10 was an assumption, and may not exactly reflect the tumor studied. Using this model, 35 Gy radiosurgery (to the 50% isodose) would be biologically equivalent to 85 Gy in 10 fractions [24] (table 1).

The increased use of stereotactic radiosurgery and stereotactic fractionated irradiation as an addition or alternative to conventional therapy for

Table 1. Animal survival after different irradiation techniques

Group	Median survival days	95% confidence interval, days	90-day survival (number of animals)	p value versus control
Control	22	21–26	0/54	–
Radiosurgery (35 Gy)	45	34–50	4/22	$< 10^{-4}$
Radiosurgery + WBRT	45	35–50	1/13	$< 10^{-4}$
WBRT	40	34–43	1/18	0.0002
85 Gy/10 fractions	49	45–51	0/16	$< 10^{-4}$
35 Gy/1 fraction	38	24–41	1/10	0.005

WBRT = 5-fraction whole-brain radiation therapy to 20 Gy.

malignant brain tumors mandates investigation into the relative effects of these approaches [27, 34]. We hypothesized that radiosurgery alone or in combination with whole-brain irradiation, would increase animal survival rates in comparison to no treatment or whole-brain irradiation alone [3, 39, 46], and that a histologic correlate could be defined with the survival response.

To identify histologic changes and effects on animal survival, we compared radiosurgery to different fractionated radiation therapy regimens including doses of calculated biologic equivalence. Such a comparison is clinically relevant, since there is an increasing use of stereotactic radiation therapy approaches for brain tumors (as an alternative to radiosurgery), despite limited knowledge regarding the number of fractions or dose necessary that might compare with radiosurgery. Rats were randomized to control (n = 54) or treatment groups after implantation of C6 glioma cells into the right frontal brain region [22, 35]. At 14 days, treated rats either had stereotactic radiosurgery (35 Gy to tumor margin; n = 22), whole-brain radiation therapy (20 Gy in 5 fractions; n = 18), radiosurgery plus whole-brain radiation therapy (n = 13), hemibrain radiation therapy (85 Gy in 10 fractions; n = 16), or single-fraction hemibrain irradiation (35 Gy; n = 10) [24].

When compared to the control group (median survival 22 days), prolonged survival was identified after radiosurgery (p < 10^{-4}), radiosurgery plus radiation therapy (p < 10^{-4}), whole-brain radiation therapy alone (p = 0.0002), hemibrain radiation therapy to 85 Gy (p < 10^{-4}), and 35 Gy hemibrain single-fraction irradiation (p = 0.005). There was no difference between the 'biologically equivalent' groups of radiosurgery and 10-fraction radiotherapy (p = 0.45), nor between radiosurgery and single-fraction nonstereotactic irradiation at the same 35-Gy margin dose (p = 0.8). Compared to the control group (mean 6.8 mm), the mean tumor diameter was reduced in all treatment groups except whole-brain radiation therapy alone. After radiosurgery, the mean tumor diameter was reduced to 5.2 ± 2.6 mm and after radiosurgery plus whole-brain irradiation, 5.3 ± 2.3 mm (fig. 2). No significant reduction in tumor size was identified after 20 Gy whole-brain irradiation alone (mean 6.3 ± 1.2 mm; $p = 0.20$). However, at the higher fractionated dose of 85 Gy, a significant reduction in tumor size was found (mean 5.4 ± 1.6 mm; $p = 0.004$), which was not different from the radiosurgery arm.

Radiosurgery ($p = 0.006$) and radiosurgery plus radiation therapy ($p = 0.009$) showed reduced tumor cell density when compared with control, a finding not observed after any fractionated regimen. Increased intratumoral edema was identified after radiosurgery ($p = 0.03$) and combined treatment ($p = 0.05$), but not after fractionated radiation therapy or 35 Gy single-fraction hemibrain irradiation. In this animal model, the addition of radiosurgery significantly increased tumor cytotoxicity, potentially at the expense of radiation effects to regional brain. We found no difference in survival benefit or tumor diameter in animals that had radiosurgery or the calculated biologically equivalent regimen of 10-fraction radiation therapy to 85 Gy. The histologic responses after radiosurgery were generally greater than those achieved with biologically equivalent doses of fractionated radiation therapy [24].

It seems that observed effects within this early 90-day period mainly are cellular responses, and not vascular responses (most of which occur later), due to the absence of vessel thrombosis or infarction. Thus, the degree of decreased tumor cellularity was proportional to observed edema (or expansion of extracellular spaces); these changes were noted only in the radiosurgery groups. These effects may represent apoptosis, necrosis, or both. It would be logical to expect a greater decrease in tumor size or even less tumor cellularity after combined radiosurgery/radiotherapy than in the other groups, since more radiation was administered. However, in the 'biologically equivalent' groups of 35-Gy radiosurgery and 10-fraction 85-Gy radiotherapy, and even when compared to the single-fraction 35-Gy arm, radiosurgery led to greater cytotoxic effects as noted by a greater reduction in cellular density. This was most likely due to the variation in the distribution of dose delivered across the tumor in radiosurgery (35 Gy at the margin, increasing to 70 Gy at the center), versus a much more uniform dose delivered across the tumor in the other regimens.

This finding suggests that dose heterogeneity within solid neoplasms may be of benefit. Such higher central doses, as well as a delayed vascular response from vessels irradiated at the tumor periphery, are likely responsible for the marked loss of central contrast enhancement often found after human acoustic tumor radiosurgery [10, 34]. Fractionated radiotherapy to the 100% isodose line did not appear to be as efficacious as radiosurgery to the 50% isodose

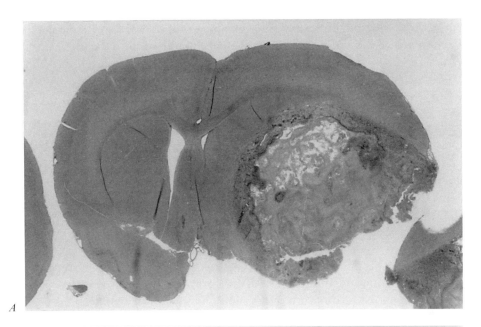

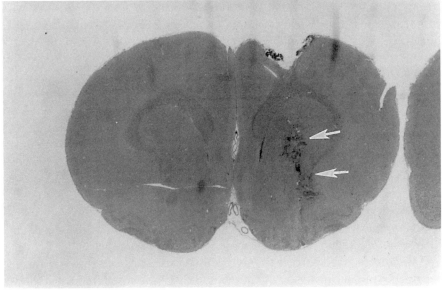

Fig. 2. A Photomicrograph (whole mount) of the rat brain 90 days after implantation of a C6 glioma irradiated with 70-Gy radiosurgery. A large tumor is evident but animal survival was prolonged significantly. *B* A different rat, also at 90 days' post-tumor implant, shows a regressed tumor after 70-Gy radiosurgery (arrows).

line. This finding might be clinically relevant to those now using stereotactic fractionated techniques, where both the benefits of precise targeting as well as the use of less homogeneous doses (at or below the 80% isodose) are being explored.

Dose Homogeneity

A careful review of the limited number of studies of how dose inhomogeneity relates to complications supports the hypothesis that dose inhomogeneity within a small radiosurgery treatment volume that matches the target volume should have little or no effect on the risk of complications. There is also support for the conclusion that dose inhomogeneity can increase the risks of complications for large target volumes treated with radiosurgery treatment volumes that are less than perfect matches for the target. The seemingly conflicting findings of the Harvard JCRT series and the University of Pittsburgh data on the relationship of dose inhomogeneity to complications seem to be explained by differences in the goodness-of-fit for the radiosurgery treatment plans. Dose inhomogeneity did not seem to be a problem with conformal multiple-isocenter Gamma knife treatment plans, while less elaborate two-isocenter Linac radiosurgery plans with possible high-dose overlap regions extending into normal tissue, were associated with a higher risk of complications. Radiosurgery treatment volumes should match the treatment volume. When this is not possible, treatment plans should be homogeneous to reduce complication risk since there is no published evidence that the higher central doses within less homogeneous radiosurgery treatment plans improve the efficacy of radiosurgery. Our knowledge of complication risk from radiosurgery is presently inadequate to know exactly how much different degrees of imperfection in matching a treatment volume to a target can be compensated for by plans with greater dose homogeneity.

The degree to which dose inhomogeneity is held to be important seems to reflect the treatment techniques available to the treating physicians and their ability to shape treatment volumes of differing homogeneity.

Because charged particle beams (proton beams) can create custom-shaped treatment volumes just as easily at the 90 or 95% isodose level as with lower isodoses like 50%, there is a natural tendency for investigators treating with this equipment to believe that dose inhomogeneity is very important for avoiding complications. Gamma knife users have historically favored treating with 50% isodose treatment volumes (the isodose percentage is always relative to the maximum dose in Gamma knife prescriptions). The greater ease of producing custom-tailored multiple isocenter treatment plans with this isodose level compared to using more homogeneous plans makes it natural for Gamma knife users to favor treating with 50% isodose volumes.

Groups using adapted linear accelerators for radiosurgery (Linac) historically have chosen a middle ground between these two extremes, usually treating with a single isocenter and prescribing to the 80% isodose level. The wider range of collimator sizes available with Linac radiosurgery makes this approach easier than with the Gamma knife. In addition, because multiple-isocenter treatment volume shaping is more time-consuming with Linac techniques than with the Gamma knife, there is a tendency to favor the greater homogeneity of a single-isocenter plan over a slightly irregular or nonspherical target over a less homogeneous, but better-fitting, multiple-isocenter plan.

In 1991, Nedzi et al. [31] reported an analysis of complications from Linac radiosurgery among 60 patients treated for brain metastasis and malignant gliomas. Their univariate analysis found that increased risks of symptomatic sequelae were significantly related or correlated with high target dose inhomogeneity as well as large treatment volume and the use of multiple isocenters. Because of these findings, they advocated treating with a single isocenter and enclosing the target with a high isodose volume whenever possible.

Benign Tumors

In the last 25 years, many patients with benign tumors have undergone stereotactic radiosurgery. Since the majority of these patients remain alive and few have had their tumors removed, little postradiosurgery tissue has been available for histologic study. We believe that the radiobiologic effect on meningiomas, schwannomas, pituitary tumors and other benign neoplasms is a combination of both cytotoxic and delayed vascular effects. To study the histologic response of benign tumors and to develop models that could be used for more in-depth molecular tests, we modified the athymic mouse subrenal capsule xenograft model for use in radiosurgery.

First, Linskey et al. [29] grafted human vestibular schwannomas into the subrenal capsule of nude mice. The model permitted accurate quantitation of small changes in tumor size, allowed for repeat measurements of tumor vascularity, and provided a high rate of graft success. The same model was used for the evaluation of human meningioma tumors. Our technique involves transplantation of the human tumor into the nude mouse within 1 h of surgical resection. After 4 weeks, the animal is reexplored to determine graft viability in situ. Only xenografts that maintain their implant size and achieve over 75% surface of vascularity are enrolled into experimental protocols that may include variable-dose stereotactic radiosurgery or a control group. For radiosurgery, animals are placed into our stereotactic frame and then imaged with plain x-rays. The tumor is located by the initial placement of a small metal coil placed beside the graft. After both vestibular schwannoma and meningioma radiosurgery, we found significant reductions in tumor volume observed after 40 Gy (within 2 weeks) and after 1 month in the 20-Gy group [29]. Similarly, tumor surface vascularity was reduced in the 20- and 40-Gy group (but not in the 10-Gy group) after 3 months of follow-up. The model proved to be an excellent technique to study the in vivo radiobiology of benign tumors after radiosurgery. We did not observe areas of intratumoral coagulative necrosis, infarction or inflammation. Thus, in this early analysis we believed that tumor size reduction was due to neoplastic cell death [20]. Current studies are being performed to evaluate growth factor production after meningioma or vestibular schwannoma radiosurgery using this xenograft model.

Some investigators have reported that apoptosis may play a significant role in both the early effects of radiosurgery for benign and malignant tumors. Since cell death may be either apoptotic or necrotic, and the temporal nature of these events different, it is important to understand how and when radiosurgery exerts an effect. Apoptosis is characterized by cell shrinkage and pyknosis without an overt inflammatory reaction. One characteristic early stage of apoptosis is deoxyribonucleic acid (DNA) cleavage. Although apoptosis may involve cell membrane or organelle effects (that later translate into DNA damage), many investigators believe that the first effect is nuclear. This cascade of events has been referred to as 'programmed cell death'. The induction of apoptosis by ionizing radiation has been evaluated in different cell lines with variable findings. Thymocytes appear to lose cell viability within hours of radiation exposure whereas some malignant cell lines require much longer intervals. In some cell lines, apoptosis occurs with the first postradiation mitosis while in others several cell divisions may occur before cell death. Tsuzuki et al. [44] theorized that the response of tumors to low-dose Gamma knife radiosurgery may be due to apoptosis since these doses would be less likely to cause vascular effects or inflammation. They evaluated expression of proliferating cell nuclear antigen (PCNA) in tumors before low-dose Gamma knife radiosurgery. Interestingly they found that all cases of malignant lymphoma showed strong positive staining for PCNA, and rapid reduction of tumor volumes after Gamma knife radiosurgery (sometimes with tumor margin doses as low as 8 Gy). They suspected that these cells received DNA damage and then rapidly entered the cell cycle leading to apoptotic death. In contrast, most benign tumors showed negative staining for PCNA and little radiographic response to low-dose radiosurgery [44].

Vascular Effects

Radiosurgery at standard clinical doses appears to inflict little injury to normal brain vessels. Available information from AVM radiosurgery or meningioma radiosurgery has shown that normal vessels rarely decrease in size or occlude after radiosurgery [47]. Since angiograms show only blood vessels > 1 mm in diameter, no comment can be made regarding the response of smaller diameter vessels using this imaging technique. Nevertheless, in our benign tumor experience, no occurrence of perforator occlusion leading to an infarct has been identified. It appears that the abnormal vessels of neoplasms or vascular malformations have a relative sensitivity to radiosurgery in comparison to normal surrounding vessels.

Radiosurgery appears to cause a proliferative vasculopathy within the blood vessels of an AVM. This process begins with endothelial cell injury [36]. Blood vessels become hyalinized, thickened and eventual luminal closure occurs. Granulation tissue may surround the AVM. This process takes many months and probably begins with an acute inflammatory reaction to radiosurgery. When this response becomes chronic, fibroblasts replace much of the mass of the AVM. Szeifert et al. [42] showed that myofibroblasts could be identified within the AVM and may provide some element of contractility to the obliteration process. Schneider et al. [36] reported a recent review of the histopathology of AVM radiosurgery from 9 specimens 10–60 months after irradiation. In most patients where histology has been obtained, only subtotal obliteration had been found (hence the need for AVM removal). We anticipate that a similar response would occur in AVMs that proceed to complete obliteration. Analysis of the complications of AVM radiosurgery [11] shows that effects in surrounding brain most likely occur from a combination of hemodynamic changes as well as parenchymal irradiation. Flickinger et al. [11] found that the 12-Gy volume surrounding the malformation was highly predictive of a symptomatic imaging change following radiosurgery. It is likely that this tissue volume has an increased sensitivity to radiation, perhaps from regional ischemia surrounding the malformation.

We believe that the occurrence of central necrosis (loss of contrast enhancement within the central portion of a tumor) after tumor radiosurgery is a vascular effect. This effect occurs in a delayed fashion, usually 3–24 months following radiosurgery. Although it may represent true cell injury rather than vessel obliteration, we believe that its time course is more consistent with delayed vessel obliteration. The effects on blood vessels play almost as important a role in the radiosurgery response as the effect on abnormal neoplastic or endothelial cells.

Lesion Generation for Functional Radiosurgery

Animal models using rat, rabbit, goat and baboon provided the basis for using radiosurgery as a lesion generator. Experiments in the 1960s showed that high radiosurgical doses (> 150 Gy) delivered to small volumes (3 × 5 mm diameter) caused consistent tissue necrosis that occurred within 1 month and did not change significantly over time [1, 19, 28]. Similar findings were identified in the rat model where doses of 150 or 200 Gy led to tissue necrosis 3 weeks after radiosurgery [21]. In 1980, Steiner et al. [40] reported an autopsy series of Gamma knife thalamotomy in the management of cancer pain and recommended that a dose of 150 Gy was necessary for the reproducible creation of a brain lesion. Later work in the rat model showed that a dose of 100 Gy caused necrosis in most (but not all) animals within 5 months and that even a dose of 50 Gy could cause complete volume necrosis in a baboon when an 8-mm collimator was used (although this finding occurred in only half the animals) [21]. Thus, the onset of the desired lesion is both dose- and volume-dependent. When this information was used in the human clinical setting, reproducible radiosurgery lesions were created using the 4-mm collimator of the Gamma knife as long as a dose of 120–200 Gy was used. When volume was increased, such as when two 4-mm isocenters were used, the volume response became less predictable and larger lesions were sometimes created [18]. As a result, consistent lesion generation is best achieved when a single 4-mm isocenter is used (or at least the smallest volume necessary for the desired clinical result). For thalamotomy or pallidotomy in humans, we have used doses of 130–140 Gy (in the thalamus) and 110 Gy (in the pallidum). With these lower doses (less than the 150 Gy proposed by Steiner et al. [40] in 1980), the desired lesion will be created, but a longer latency may be necessary (1–3 months) (fig. 3). Histologically, the radiosurgery lesion consists of coagulative necrosis within the target volume, a thin gliotic rim, and rapid normalization of parenchyma within 1–2 mm [20, 21]. Although imaging studies may show long TR signal changes in white matter tracts surrounding the lesion, these are invariably asymptomatic.

Pharmacologic Radioprotection for Radiosurgery

Future improvements in the results of stereotactic radiosurgery will be related to better patient selection, dose planning, radiosensitization of the target [8], and possibly protection of the brain surrounding the target. Prior investigations into radiation protection have examined myelin and lipid effects, as well as vascular effects [32, 33]. We previously investigated the potential

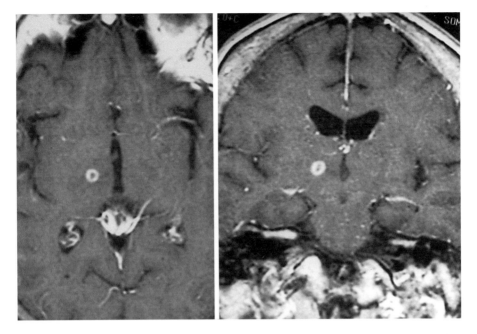

Fig. 3. Axial and coronal MR scans (short TR with contrast) 2 months after a right thalamotomy (140 Gy) for essential tremor. This patient had significant concomitant medical illnesses. A 4-mm discrete radionecrotic lesion was identified.

radioprotectant effects of high-dose pentobarbital; however we detected no specific benefit in our radiosurgical model [23] (fig. 4). Bernstein et al. [4] reported that the 21-aminosteroid U-74389G reduced brachytherapy-induced brain injury in a rat model and Buatti et al. [7] reported that U-74389G provided radiation protection in a model where cats received either 21-amino-steroid or corticosteroid. 21-Aminosteroids may provide protection against brain radiation injury by inhibition of lipid peroxidation and a selective action on vascular endothelium [2, 5, 6, 12]. We hypothesized that the 21-aminosteroid U-74389G would reduce radiosurgery-related brain injury without attenuating the target volume response. Kallfass et al. [17] using a celiac artery irradiation model, demonstrated inflammatory vascular effects within 24 h. As a membrane stabilizer, 21-aminosteroids block the release of free arachidonic acid from injured cell membranes [12]. The pharmacologic prevention of this early response might prevent a reactive cascade that would otherwise end in a chronic radiation vasculopathy.

Hornsey et al. [16] studied the effects of vasoactive drugs such as dipyr-amidole and desferrioxamine, and found reduction of spinal cord radiation

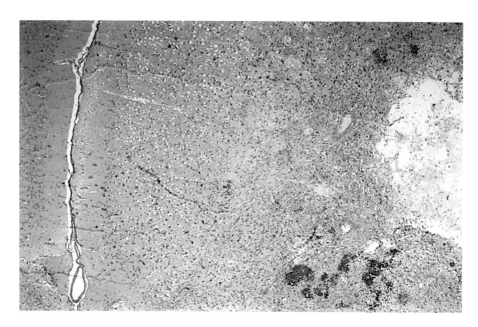

Fig. 4. Photomicrograph of the rat brain 150 days after 100-Gy radiosurgery following administration of pentobarbital. Radiation necrosis was created (right side of image) and was surrounded by edema and vasculopathy. Note the mass effect on the interhemispheric fissure.

damage. They postulated that the beneficial effect was due to improved spinal cord blood flow. Whether an agent selective to blood vessels could improve regional blood flow or provide regional vascular stability with limitation of regional edema remains to be identified. Braughler et al. [6] reported that the effects of 21-aminosteroid were 100 times more potent than desferrioxamine. In a separate study using cultured bovine endothelial cells, Audus et al. [2] showed that 21-aminosteroids associate with the hydrophobic segments of endothelial cell membranes, and thus can exert their action on the local cerebrovasculature. Oxygen radical initiated peroxidation of vascular membranes is catalyzed by free iron release from hemoglobin, ferritin and transferrin. If not prevented, lipid peroxidation progresses over the surface of the cell membrane to cause disruption of phospholipid-dependent enzymes, ionic gradients, and later membrane lysis. Repair of these effects will be manifest as a radiation-induced vasculopathy. U-74389G as a lipid antioxidant and free-radical scavenger was found in our study to limit radiation-induced vessel changes and prevent regional edema [25].

We placed rats into one of four experimental groups before radiosurgery: control (n = 47), low-dose U-74389G (5 mg/kg; n = 30); high-dose U-74389G

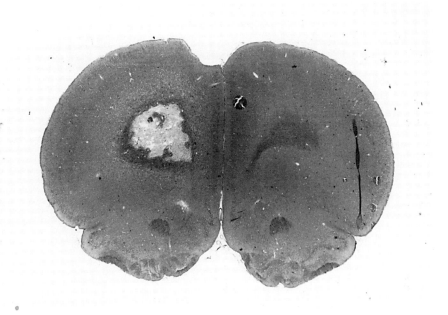

Fig. 5. Whole mount of the rat brain after 100-Gy radiosurgery with administration of U-74389G (15 mg/kg) shows necrosis with minimal peripheral vasculopathy and no regional mass effect.

(15 mg/kg; n = 20), and methylprednisolone (2 mg/kg; n = 48). The drug was administered 1 hr before radiosurgery (4-mm Gamma knife collimator) of the normal rat frontal lobe (single-fraction maximum doses of 50, 100 or 150 Gy). We studied the parenchymal response at 90 or 150 days for diameter of necrosis, and the findings of radiation-induced vasculopathy, brain edema and gliosis (fig. 5).

At 50 Gy, no animal developed histologic changes whereas all 150-Gy animals developed radiation necrosis without drug-induced protection from vascular changes or edema. In 100-Gy animals, high-dose aminosteroid reduced radiation-induced vasculopathy at 90 days (p = 0.06) and at 150 days (p = 0.02), and reduced regional edema at 90 days (p = 0.01) and at 150 days (p = 0.03). Low-dose aminosteroid and corticosteroid provided no protection [25].

The 21-aminosteroid U-74389G provided protection after a single intravenous 15-mg/kg dose against 100-Gy radiation-induced vasculopathy and edema. High-dose 21-aminosteroids appear to have optimal properties for radiosurgery: surrounding brain protection without reducing the therapeutic effect desired within the target volume. We found a dose-response relationship for prevention of vascular effects, and that this likely translates into prevention from the development of regional cerebral edema.

Ideally, a pharmacologic radioprotection agent would not impact on the treatment delivered to the brain lesion, but would protect the surrounding brain in the region of dose fall-off. In this study, only protection of normal brain tissue and not the potential protection of an abnormal target was evaluated. We are performing experiments using our malignant glioma radiosurgery model to determine whether U-74389G prevents the clinical benefit of radiosurgery on a tumor cell line [22]. If no lesional protectant effect is identified, we plan to initiate a randomized clinical trial for human tumor radiosurgery.

References

1   Andersson B, Larsson B, Leksell L, et al: Histopathology of late local radiolesions in the goat brain. Acta Radiol Ther Phys Biol 1970;9:385–394.
2   Audus KL, Guillot FL, Braughler JM: Evidence for 21-aminosteroid association with the hydrophobic domains of brain microvessel endothelial cells. Free Radic Biol Med 1991;11:361–371.
3   Barker M, Deen DF, Baker DG: BCNU and x-ray therapy of intracerebral 9L rat tumors. Int J Radiat Oncol Biol Phys 1979;5:1581–1583.
4   Bernstein M, Ginsberg H, Glen J: Protection of iodine-125 brachytherapy brain injury in the rat with the 21-aminosteroid U-74389G. Neurosurgery 1992;31:923–928.
5   Braughler JM: Lipid peroxidation-induced inhibition of gamma-aminobutyric acid uptake in rat brain synaptosomes: Protection by glucocorticoids. J Neurochem 1985;44:1282–1288.
6   Braughler JM, Pregenzer JF, Chase RL, Duncan LA, Jacobsen EJ, McCall JM: Novel 21-aminosteroids as potent inhibitors of iron-dependent lipid peroxidation. J Biol Chem 1987;262:10438–10440.
7   Buatti JM, Friedman WA, Theele DP, Bova FJ, Mendenhall WM: The lazaroid U-74389G protects normal brain from stereotactic radiosurgery-induced radiation injury. Int J Radiat Oncol Biol Phys 1996;34:591–597.
8   Cohen JD, Robins HI, Javid MJ: Radiosensitization of C6 glioma by thymidine and 41.8 °C hyperthermia. J Neurosurg 1990;72:782–785.
9   Flickinger J, Kondziolka D, Lunsford LD, et al: A multi-institutional experience with stereotactic radiosurgery for solitary brain metastases. Int J Radiat Oncol Biol Phys 1994;28:797–802.
10 Flickinger JC, Kondziolka D, Pollock B, et al: Evolution of technique for vestibular schwannoma radiosurgery and effect on outcome. Int J Radiat Oncol Biol Phys 1996;36:275–280.
11 Flickinger JC, Kondziolka D, Pollock B, et al: Complications from arteriovenous malformation radiosurgery: Multivariate analysis and modeling. Int J Radiat Oncol Biol Phys 1997;38:485–490.
12 Hall ED, Travis MA: Inhibition of arachidonic acid-induced vasogenic brain edema by the non-glucocorticoid 21-aminosteroid U-74006F. Brain Res 1988;451:350–352.
13 Hall EJ, Brenner DJ: The radiobiology of radiosurgery: Rationale for different treatment regimes for AVMs and malignancies. Int J Radiat Oncol Biol Phys 1993;25:381–385.
14 Henderson SD, Kimler BF, Morantz RA: Radiation therapy of 9L rat brain tumors. Int J Radiat Oncol Biol Phys 1981;7:497–502.
15 Hornsey S, Morris CC, Myers R: The relationship between fractionation and total dose for x-ray induced brain damage. Int J Radiat Oncol Biol Phys 1981;7:393–396.
16 Hornsey S, Myers S, Jenkinson T: The reduction of radiation damage to the spinal cord by post-irradiation administration of vasoactive drugs. Int J Radiat Oncol Biol Phys 1990;18:1437–1442.
17 Kallfass E, Kramling HJ, Schultz-Hector S: Early inflammatory reaction of the rabbit coeliac artery wall after combined intraoperative and external irradiation. Radiother Oncol 1996;39:167–178.

18 Kihlstrom L, Guo W, Lindquist C, et al: Radiobiology of radiosurgery for refractory anxiety disorders. Neurosurgery 1995;36:294–302.

19 Kihlstrom L, Hindmarsh T, Lax I, et al: Radiosurgical lesions in the normal human brain 17 years after gamma knife capsulotomy. Neurosurgery 1997;41:396–402.

20 Kondziolka D, Linskey ME, Lunsford LD: Animal models in radiosurgery; in Alexander E, Loeffler J, Lunsford LD (eds): Stereotactic Radiosurgery. New York, McGraw-Hill, 1993, pp 51–64.

21 Kondziolka D, Lunsford LD, Claassen D, et al: Radiobiology of radiosurgery. 1. The normal rat brain model. Neurosurgery 1992;31:271–279.

22 Kondziolka D, Lunsford LD, Claassen D, et al: Radiobiology of radiosurgery. II. The rat C6 glioma model. Neurosurgery 1992;31:280–288.

23 Kondziolka D, Somaza S, Flickinger JC, et al: Cerebral radioprotective effects of high-dose pentobarbital evaluated in an animal radiosurgery model. Neurol Res 1994;16:456–459.

24 Kondziolka D, Somaza S, Comey C, Lunsford LD, Claassen D, Pandalai S, Maitz A, Flickinger JC: Radiosurgery and fractionated radiation therapy: Comparison of different techniques in an in vivo rat glioma model. J Neurosurg 1996;84:1033–1038.

25 Kondziolka D, Somaza S, Martinez AJ, et al: Radioprotective effects of the 21-aminosteroid U-74389G for stereotactic radiosurgery. Neurosurgery 1997;41:203–208.

26 Larson DA: Radiosurgery and fractionation; in Kondziolka D (ed): Radiosurgery 1995. Radiosurgery. Basel, Karger, 1996, vol 1, pp 261–267.

27 Larson DA, Flickinger JC, Loeffler JS: The radiobiology of radiosurgery. Int J Radiat Oncol Biol Phys 1993;25:557–561.

28 Larsson B, Leksell L, Rexed B, et al: The high-energy proton beam as a neurosurgical tool. Nature 1958;182:1222–1223.

29 Linskey ME, Martinez AJ, Kondziolka D, et al: The radiobiology of human acoustic schwannoma xenografts after stereotactic radiosurgery evaluated in the subrenal capsule of athymic mice. J Neurosurg 1993;78:645–653.

30 Malaise EP, Fertil B, Chavaudra N, et al: Distribution of radiation sensitivities for human tumor cells of specific histological types: Comparison of in vitro to in vivo data. Int J Radiat Oncol Biol Phys 1986;12:617–624.

31 Nedzi L, Kooy H, Alexander E, et al: Variables associated with the development of complications from radiosurgery of intracranial tumors. Int J Radiat Biol Phys 1991;21:591–599.

32 Oldfield EH, Friedman R, Kinsella T, Moquin R, Olson JJ, Orr K, Deluca AM: Reduction in radiation-induced brain injury by use of pentobarbital or lidocaine protection. J Neurosurg 1990; 72:737–744.

33 Olson JJ, Friedman R, Orr K, Delaney T, Oldfield EH: Cerebral radioprotection by pentobarbital: Dose-response characteristics and association with GABA agonist activity. J Neurosurg 1990;72: 749–758.

34 Pollock BE, Lunsford LD, Kondziolka D, et al: Outcome analysis of acoustic neuroma management: A comparison of microsurgery and stereotactic radiosurgery. Neurosurgery 1995;36:215–229.

35 San-Galli F, Vrignaud P, Robert J, et al: Assessment of the experimental model of transplanted C6 glioblastoma in Wistar rats. J Neurooncol 1989;7:299–304.

36 Schneider BF, Eberhard DA, Steiner LE: Histopathology of arteriovenous malformations after gamma knife radiosurgery. J Neurosurg 1997;87:352–357.

37 Schwachenwald R, Engebraten O, Valen H, et al: A technique for studying single-dose radiation effects on glioma invasiveness in tissue culture – A pilot study; in Steiner L (ed): Radiosurgery: Baseline and Trends. New York, Raven Press, 1992, pp 101–109.

38 Shaw E, Scott C, Souhami L, et al: Radiosurgery for the treatment of previously irradiated recurrent primary brain tumors and brain metastases: Initial Report of Radiation Therapy Oncology Group Protocol 90-05. Int J Radiat Oncol Biol Phys 1996;34:647–654.

39 Steinbok P, Mahaley MS, U R, et al: Treatment of autochthonous rat brain tumors with fractionated radiotherapy: The effects of graded radiation doses and of combined therapy with BCNU or steroids. J Neurosurg 1980;53:68–72.

40 Steiner LE, Forster D, Leksell L, et al: Gammathalamotomy in intractable pain. Acta Neurochir 1990;52:173–184.

41   Stuschke M, Budach V, Sack H: Radioresponsiveness of human glioma, sarcoma, and breast cancer spheroids depends on tumor differentiation. Int J Radiat Oncol Biol Phys 1993;27:627–636.

42   Szeifert G, Kemeny AA, Timperley W, et al: The potential role of myofibroblasts in the obliteration of arteriovenous malformations after radiosurgery. Neurosurgery 1997;40:61–66.

43   Thompson BG, Coffey RJ, Flickinger J, et al: Stereotactic radiosurgery of small intracranial tumors: Neuropathological correlation in three patients. Surg Neurol 1990;33:96–104.

44   Tsuzuki T, Tsunoda S, Sakaki T, et al: Tumor cell proliferation and apoptosis associated with the Gamma Knife effect. Stereotact Funct Neurosurg 1996;66(suppl 1):39–48.

45   van der Kogel A: Central nervous system radiation injury in animal models; in Gutin P, Leibel S, Sheline G (eds): Radiation Injury to the Nervous System. New York, Raven Press, 1991, pp 91–111.

46   Wheeler KT, Kaufman K: Influence of fractionation schedules on the response of a rat brain tumor to therapy with BCNU and radiation. Int J Radiat Oncol Biol Phys 1980;6:845–849.

47   Yamamoto M, Jimbo M, Kobayashi M, et al: Long-term results of radiosurgery for arteriovenous malformation: Neurodiagnostic imaging and histological studies of angiographically confirmed nidus obliteration. Surg Neurol 1992;37:219–230.

48   Yang X, Darling JL, McMillan TJ, et al: Radiosensitivity, recovery and dose-rate effect in three human glioma cell lines. Radiother Oncol 1990;19:49–56.

Douglas Kondziolka, MD, University of Pittsburgh Medical Center,
Suite B-400, Department of Neurological Surgery,
200 Lothrop Street, Pittsburgh, PA 15213 (USA)
Tel. (412) 647 6782, Fax (412) 647 0989

Lunsford LD, Kondziolka D, Flickinger JC (eds): Gamma Knife Brain Surgery.
Prog Neurol Surg. Basel, Karger, 1998, vol 14, pp 39–50

Principles of Dose Prescription and Risk Minimization for Radiosurgery

John C. Flickinger[a,b], *Douglas Kondziolka*[b,a], *L. Dade Lunsford*[b,a,c]

Departments of [a]Radiation Oncology, [b]Neurological Surgery, and
[c]Radiology, University of Pittsburgh School of Medicine, Pittsburgh, Pa., USA

Basic Principles

Dose prescription in radiosurgery has some similarities with conventional fractionated radiotherapy but also several important differences. One similarity is the concept of paired dose-response curves with desired outcome (tumor control, arteriovenous malformation (AVM) obliteration, etc.) side by side with complications (such as radiation necrosis). Figure 1 illustrates this concept which is used in introductory texts for radiation oncologists [13]. The relationship of successful outcome versus complications at different doses is determined by the shapes and separation of the two dose-response curves. The choice of an optimum dose is based on the balance of these two outcomes.

In the hypothetical example shown in figure 1, the rates of successful outcome versus complications for 15 Gy are 64 vs. 7%, for 20 Gy are 91 vs. 37%, and for 25 Gy are 98 vs. 76%, respectively. The optimum dose for treating any indication would depend on the consequences of not achieving the desired outcome (such as tumor control) and the severity of the complication. For an AVM, a 64% obliteration rate with a 7% risk of radiation necrosis might be the best of these three choices. This is because failure to achieve obliteration can be addressed with repeat radiosurgery with reasonable risks of morbidity from retreatment or from hemorrhage occurring before obliteration is achieved. If the desired outcome is cure of a malignant primary tumor with failure to achieve this outcome resulting in a rapid death from tumor progression, then a higher dose such as 15 or 20 Gy would be desirable in this example, if the severity of the complication represented is not too great. If the complication is temporary neurological deterioration but the risks of severe permanent sequelae are low, then a higher complication rate may be acceptable.

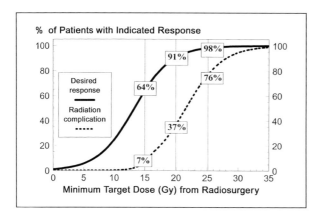

% of Patients with Indicated Response

Fig. 1. Theoretical paired sigmoid dose-response curves for desired outcome (such as tumor control or vascular malformation obliteration) and for complications.

Inadequacy of the Traditional Paired Dose-Response Model

While the pair of sigmoid dose-response curves shown in Figure 1 is useful for illustrating certain concepts of the balance between desired outcome versus complication risk in fractionated radiotherapy, it doses not accurately represent the balance of these two endpoints in radiosurgery. While the effect of treatment volume upon complication-risk is modest for large-field fractionated radiotherapy and can usually be ignored, complications depend upon volume more than anything else within the range of doses used and volumes treated in radiosurgery. For radiosurgery, the single dose-response curve for complications shown in figure 1 is inadequate, since a whole family of complication curves is needed for different treatment volumes and because complication dose-response function for each volume are less steep, they vary less with dose than figure 1 illustrates.

Our initial studies of complication risk from radiosurgery correlated morbidity risk with volume for AVMs and with nerve length irradiated for acoustic neuromas but could not identify any correlation with either minimum target dose (D_{min}) or maximum dose (D_{max}) within the range of doses prescribed [2, 8–10, 15]. The relatively small contributions of the differences in doses prescribed to radiosurgery complications were overshadowed even in multivariate analysis by the tremendous volume effects. These effects were only detectable later in the analysis of a much larger series of patients [3, 5–7].

Another volume-related factor that is important in radiosurgery is the quality of the treatment plan. The inherent goal of radiosurgery is to administer a sharply-defined radiation treatment volume that perfectly matches the target tissue, thereby sparing irradiation of surrounding tissue as much as possible. The quality of a radiosurgery plan is how well its chosen treatment isodose

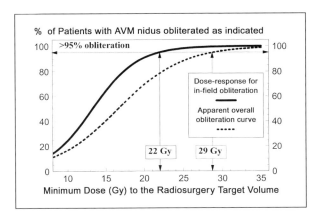

Fig. 2. Comparison of the true dose-response curve for in-field AVM obliteration (solid curve) to the apparent dose-response curve for overall obliteration of AVM nidus (dashed curve) according to minimum dose (D_{min}) administered within the radiosurgery treatment volume. The curves were derived from analysis of 197 AVM patients with 3-year angiographic follow-up after radiosurgery at the University of Pittsburgh [11].

matches the target. When normal tissue is included within the radiosurgery treatment volume, the risk of complications increases to a degree dependent on the volume of normal tissue included and the dose administered. On the other hand, failing to include part of the target within the treatment volume can lead to failure to achieve the desired outcome such as tumor control or AVM obliteration, depending on how much the unenclosed portion is underdosed.

For various reasons, clinicians frequently fall short of the goal of a perfect-fitting radiosurgery treatment volume to various degrees. Sometimes this goal is not achieved because of imaging inadequacies. We correlated reduced risks of complications after acoustic neuroma radiosurgery with better quality treatment plans from improvements in imaging (switching from CT to high-resolution MR imaging) and more elaborate multiple isocenter treatment plans (made easier to construct because of improved planning software) [6]. Problems with control of cerebellar brain metastases were identified in the Harvard Joint Center experience that may be attributed to inadequate imaging of the posterior fossa because of CT artifact [17]. Our analysis of AVM obliteration after radiosurgery found that inadequate target definition was the primary cause of treatment failure [7].

Dose-response curves derived from outcome data that do not account for the possibility of inadequate target definition can be misleading. Figure 2 compares a misleading dose-response curve for overall AVM obliteration from

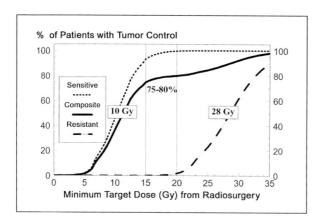

Fig. 3. Composite dose-response curve (solid line) constructed for a mixed population consisting of relatively radiosensitive tumors in 80% of patients and relatively radioresistant tumors in 20% of patients. The individual dose-response curves for the radiosensitive and radioresistant tumors are also shown.

the University of Pittsburgh experience to the true dose-response function for in-field AVM obliteration [7]. If the overall obliteration rate were presented by itself, it could mislead the reader into believing that a dose ≥ 29 Gy is necessary to ensure AVM obliteration with $> 95\%$ certainty, when only 22 Gy is needed.

A final inadequacy of the paired sigmoid dose-response curves pictured in figure 1 for radiosurgery (and for fractionated radiotherapy), is the use of a relatively sharp dose-response curve to represent tumors (the response function in figure 1 is actually for AVM obliteration). While radiobiology studies of uniform tumor cell lines in animal and cell culture models show similar dose-response functions, this does not accurately represent most tumors encountered in the clinic, where it is difficult to show any dose response at all within the range of doses commonly used in radiosurgery (or in conventional radiotherapy). While there is only a gradual improvement in tumor control with increasing dose on the plateau portion of the curve above 90%, many tumors have relatively flat dose responses with tumor control rates closer to 50%, which should be the steepest part of the dose-response curve. Tumor heterogeneity can explain the phenomenon of a flat dose response.

Figure 3 illustrates a relatively flat dose response between doses of 15–20 Gy for a heterogeneous tumor population. Eighty percent of the composite population consists of patients with relatively radiosensitive tumors with a traditional sigmoid-shaped dose-response curve that show 50% control after 10 Gy. The other 20% of patients with relatively radioresistant tumors are pathologically indistinguishable from the sensitive tumors but have a different dose-response curve and require a dose of 28 Gy for 50% tumor control. The composite dose-response function is relatively flat within the limited range of therapeutic doses of 15–20 Gy explored clinically. A small improvement in tumor control from 75% at 15 Gy to 80% at 20 Gy can be undetectable even in a large clinical series. Small differences in tumor control with various doses can be obscured by recurrences because of inadequate targeting or uncertainties in distinguishing between tumor recurrence and radiation injury.

Because of these reasons, we have no absolute and reliable tumor dose-response curves to guide radiosurgical dose prescription. Dose responses are difficult to define with certainty in patients with malignant primary and metastatic brain tumors because of the difficulty distinguishing between tumor progression versus necrosis during patients' limited lifespans and the added complexities of other treatment such as surgical resection, and fractionated radiotherapy. Benign tumors are also difficult to study since their natural history requires long-term follow-up. Marginal or out-of-field recurrences need to be separated from in-field recurrences in patients who have undergone previous surgical resections.

We looked for a dose response in our acoustic neuroma experience, where we lowered our average minimum tumor prescription dose from 18 to 14 Gy over time [6]. No difference in tumor control related to dose is yet evident. We will continue to reanalyze this in the future to make sure that no difference arises with longer follow-up. We have not found any dose response in our meningioma experience either, although this tumor is more difficult to study because of the higher rate of prior surgery and the possibility that surgery could have left microscopic disease outside the radiosurgery target. It appears that 80–90% of all benign tumors treatable by radiosurgery are controlled with doses of 12–18 Gy. We were similarly unable to define a clear dose-response relationship for radiosurgery of malignant tumor. In the gamma knife users study of solitary brain metastases, we found poorer local control with radiosurgery alone (50% with a median dose 17 Gy) compared to radiosurgery (85% with a median dose 16 Gy for radiosurgery) combined with whole brain radiosurgery (30 Gy) [4]. Because others have reported high local control rates with radiosurgery alone to 20 Gy, we use this higher dose for radiosurgery alone as allowed by brain tolerance.

Dose-Volume Relationships for Complications

Over the last 10 years, dose prescriptions for radiosurgery have been based upon volume-dependent projected risks of radiation necrosis to brain tissue. Risk predictions from Kjellberg's 1% isoeffect line and from the integrated logistic formula still serve as the standard guidelines for dose prescription in radiosurgery [1, 14, 16]. Kjellberg's 1% isoeffect line was empirically drawn

Table 1. Comparison of dose-volume radiosurgery guidelines from Kjellberg's 1% isoeffect line (1% dose) and the integrated logistic formula's 3% radiation necrosis risk predictions (IL 3%)

Diameter, mm	Volume, cc	1% dose, Gy	IL 3%, Gy
12.5	1.02	27.5	34.0
15.0	1.77	25.0	29.0
17.5	2.81	22.5	23.0
20.0	4.19	20.0	18.0
22.5	5.96	18.7	16.5
25.0	8.18	17.5	14.5
27.5	10.89	16.5	13.5
30.0	14.14	15.0	13.0

from an amalgamation of a limited clinical experience with proton beam irradiation and from animal experiments with small proton and photon beams. Kjellberg's own data shows enough complications below the 1% line to indicate it underestimates complication risk. The integrated logistic formula is a mathematical projection of brain necrosis risks from large- and small-field radiotherapy in humans and animals calculated over inhomogeneous radiation dose distributions used for radiosurgery. Table 1 compares the relatively similar dose-volume guidelines from Kjellberg's 1% isoeffect line and the integrated logistic formula's 3% risk guidelines.

Neither Kjellberg's 1% isoeffect line nor the integrated logistic formula predict temporary postradiosurgery imaging (PRI) changes, which in many patients may be asymptomatic [7, 8]. PRI are more related to the total volume of tissue irradiated to high doses (including the target) rather than just the dose distribution to surrounding normal tissue which is used to calculate the projected risk of permanent injury (radiation necrosis) with the integrated logistic formula [7].

We evaluated PRI changes that developed in 57 of 277 patients with AVM or benign tumors (meningioma or acoustic neuroma) from 1 to 23 months after radiosurgery [2, 9]. PRI changes developed in a greater proportion of AVM vs. tumor patients (31 vs. 8%). This difference should not occur unless tissue within the target contributed to this injury response. It also indicated that radiosurgery sequelae of tumors and AVMs should be studied separately.

We subsequently analyzed 307 AVM patients who received gamma knife radiosurgery at the University of Pittsburgh between 1987 and 1993 with regular clinical and imaging follow-up from a minimum of 2 to a maximum

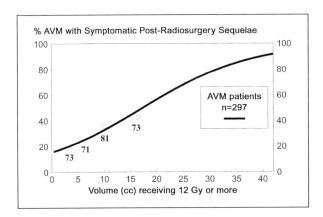

Fig. 4. Risk prediction curves for AVM patients that correlate 12-Gy volume to risks for developing all (symptomatic and asymptomatic) PRI changes. The numbers next to the four boxes represent the number of patients in each data quartile. The position of each box indicates the percentage of complications versus the mean 12-Gy volume for the quartile.

of 8 years [7]. PRI changes (including both symptomatic or asymptomatic) developed in 30.5% of patients. Multivariate logistic regression analysis found that the total volume of tissue (including the AVM target) that received 12 Gy or more (the 12-Gy volume) was the only independent variable that correlated significantly with PRI changes (p < 0.0001). Other factors including dose rate, D_{max}, D_{min}, treatment isodose, volume, target dose inhomogeneity, and the number of isocenters were not independently significant in this study or in a subsequent reanalysis. To some degree, D_{min} and volume are nonetheless important since the 12-Gy volume is a parameter that correlates with both the volume of the target being irradiated and the dose administered. Figure 4 shows the 12-Gy volume risk-prediction curves for the development of any PRI changes from this study.

Symptomatic PRI changes developed in 10.7% of all the AVM patients at 7 years (with no cases developing more than 2 years after treatment) [7]. PRI changes resolved within 3 years significantly less often (p = 0.0274) in patients with symptoms (53%) compared to asymptomatic patients (95%). The 7-year actuarial rate for developing persistent symptomatic PRI changes (adverse radiation sequelae) in the series was 5.0%.

A subsequent reanalysis that included a total of 332 AVM patients concentrated on identifying factors affecting the risk of symptomatic PRI changes [5]. By evaluating the relative risks of symptomatic sequelae for different locations, we constructed a postradiosurgery injury expression (PIE) score for AVM location (table 2). Multivariate logistic regression analysis of sympto-

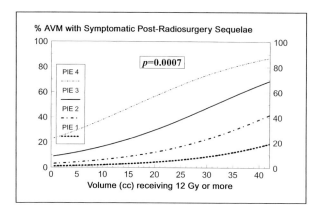

Fig. 5. Risk prediction curves derived from 332 AVM patients that correlate 12-Gy volume to risks for developing symptomatic post-radiosurgery sequelae according to PIE score.

Table 2. PIE score for AVM location: the risk of symptomatic sequelae increases with PIE score

PIE score	Location
1	Frontal lobe
2	Cerebellum, temporal lobe, or parietal lobe
3	Occipital lobe or basal ganglia
4	Medulla, thalamus, intraventricular, pons, or corpus callosum

matic postradiosurgery sequelae identified independent significant correlations with PIE location score (p = 0.0007) and 12-Gy volume (p = 0.008) but none of the other factors tested (p ≥ 0.3) including the addition of MR targeting, average radiation dose in 20 cm³, prior hemorrhage or neurological deficit. Figure 5 shows the risk of symptomatic sequelae predicted from PIE score and 12-Gy volume. The PIE score is a way of categorizing radiosurgery sequelae that is based upon how likely symptoms will be evident from a radiosurgery injury response in that portion of the brain rather than some inherent difference in radiosensitivity.

The risk of radiation necrosis (symptomatic sequelae persisting for more than 2 years) was significantly correlated with PIE score (p ≤ 0.048) but not 12-Gy volume [5]. We discovered that symptomatic postradiosurgery sequelae were more likely to resolve in patients treated with minimum AVM nidus doses (D_{min}) < 20 Gy compared to D_{min} ≥ 20 Gy (89 vs. 36%, respectively; p = 0.006) [5]. Fortunately for the patients, there were too few with persistent symptomatic radiation injury to construct a reliable risk prediction model for this endpoint. We recently started a multicenter retrospective analysis of symptomatic sequelae in AVM patients following gamma knife radiosurgery to better define factors that distinguish temporary from permanent injury. We expect this study will produce a reliable model for estimating the risk of symptomatic radiation necrosis, the complication endpoint of greatest importance to clinicians and patients.

An RTOG phase 1 dose-escalation study of radiosurgery in recurrent primary (n = 38) and metastatic (n = 64) brain tumors was recently reported [19]. Unacceptable toxicity was defined as irreversible RTOG Grade 3, or any Grade 4–5, CNS toxicity in >20% of patients per treatment arm within 3 months of radiosurgery. For tumors <20 mm in diameter, 0/40 patients developed ≥ Grade 3 CNS toxicity with D_{min} of 18, 21 and 24 Gy. For tumors 21–30 mm, D_{min} of 15, 18 and 21 Gy were tested. Two out of 42 of these patients developed Grade 3 CNS toxicity with doses of 15 and 21 Gy. More serious toxicity was encountered for patients with 31–40-mm diameter tumors. At the dose level of 12 Gy, 1/21 developed Grade 5 toxicity. At 15 Gy, 0/22 patients had Grade 3 or greater toxicity, but at 18 Gy, 4/18 patients developed significant toxicity (2 Grade 3 and 2 with Grade 4). Because half of all postradiosurgery injury responses in long-term survivors occur from 6 months to 2 years after radiosurgery, we do not feel that this study adequately addresses long-term toxicity. We advocate continuing to follow traditional dose-volume guidelines from the integrated logistic formula or the Kjellberg 1% isoeffect line because of the adequate tumor control and limited toxicity reported in series using these guidelines.

Cranial Nerve Tolerance

The risk of cranial nerve injury is another factor that the dose-prescription guidelines of Kjellberg or the integrated logistic formula do not address. Special sensory nerves (the optic and auditory nerves) appear to be the most sensitive to radiation injury, followed by somatic sensory and finally motor nerves [3]. Acoustic neuromas (vestibular schwannomas) provide an excellent model to study the effects of radiosurgery on three different cranial nerves. We found that the risks of facial, trigeminal, and auditory neuropathies after radiosurgery correlate closely with the length of nerve irradiated [3, 15]. Since the nerve length irradiated depends upon the size of the tumor irradiated and the quality of the dose plan, the best strategies to limit postradiosurgery neuropathies in the treatment of acoustic neuroma are to irradiate with a properly-fitting radiosurgery treatment volume while the tumor is small, rather than observe it until it becomes larger and riskier to treat. We discerned dose/nerve-length response curves from the University of Pittsburgh acoustic neuroma data to

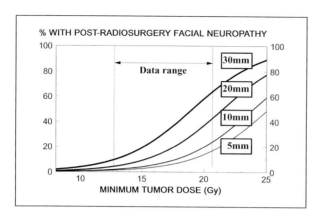

Fig. 6. Dose-response curves for temporary or permanent facial neuropathy developing after acoustic neuroma radiosurgery derived from analysis of 238 patients. The labels on each curve indicate the combined transverse tumor diameter (intra- plus extracanalicular diameters).

predict facial weakness, numbness, and hearing loss after radiosurgery [3]. Figure 6 shows one family of dose-response curves for facial neuropathy. We do not routinely consult these risk-prediction curves for help in dose prescription. Our current policy is to treat acoustic tumors to a minimum tumor dose of 12–14 Gy depending upon the size of the tumor and how important the goals of hearing preservation and avoidance of facial or trigeminal neuropathies are in each individual patient. Higher doses may even be considered in patients with tumors that recur following prior resection who have no hearing and facial weakness and/or numbness. Although we have not detected any decrease in tumor control in lowering our average treatment doses (D_{min}) from 18 to 12–14 Gy, longer follow-up is necessary to detect any decrease in long-term tumor control with dose reduction [6]. Because of the relatively low morbidity of acoustic neuroma radiosurgery with the current technique, we are presently reluctant to lower treatment doses further out of fear of compromising tumor control.

Because of the importance of vision, the dose administered to optic nerves and/or chiasm is what frequently determines the minimum treatment volume dose for radiosurgery. Radiation dose to the optic apparatus can be limited when treating nearby targets by selective beam channel blocking with gamma knife units or with modified arc weighting/positioning. The tolerance of the optic apparatus and cavernous sinus nerves was analyzed in the combined early radiosurgery experience from the Harvard Joint Center and the University of Pittsburgh [20]. We identified a dose response for optic nerve/chiasm injury

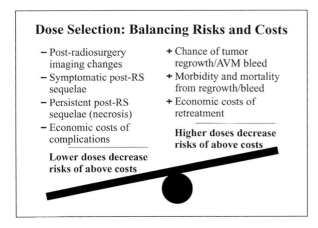

Dose Selection: Balancing Risks and Costs

– Post-radiosurgery
imaging changes
– Symptomatic post-RS
sequelae
– Persistent post-RS
sequelae (necrosis)
– Economic costs of
complications

+ Chance of tumor
regrowth/AVM bleed
+ Morbidity and mortality
from regrowth/bleed
+ Economic costs of
retreatment

**Higher doses decrease
risks of above costs**

**Lower doses decrease
risks of above costs**

Fig. 7. Illustration of the competing risks associated with choosing higher versus lower doses for radiosurgery.

but not for the more radioresistant nerves in the cavernous. Because the lowest dose that optic neuropathy developed was 9.75 Gy, we recommended limiting optic nerve/chiasm doses to a maximum of 8 Gy.

Risk Decision Analysis

The preceding discussion outlines most of what is known and what is yet to be defined concerning dose-response functions for both the desired outcome and complications with radiosurgery for various indications. Even when all of these dose-response functions become perfectly defined, it will still be difficult to decide on the optimum radiosurgical dose for each patient. Figure 7 illustrates the concept of competing risks changing with dose in radiosurgery. The choice of the optimum dose depends upon how the physician and patient value or weigh the various possible outcomes.

References

1   Flickinger JC: The integrated logistic formula and prediction of complications from radiosurgery. Int J Radiat Oncol Biol Phys 1089;17:879–885.
2   Flickinger JC, Kondziolka D, Kalend AM, Maitz AH, Lunsford LD: Radiosurgery-related imaging changes in surrounding brain: Multivariate analysis and model evaluation; in Kondziolka D (ed): Radiosurgery 1995. Radiosurgery. Basel, Karger, 1996, vol 1, pp 229–236.
3   Flickinger JC, Kondziolka D, Lunsford LD: Dose and diameter relationships for facial, trigeminal, and acoustic neuropathies following acoustic neuroma radiosurgery. Radiother Oncol 1996;41: 215–219.

4   Flickinger JC, Kondziolka D, Lunsford LD, et al: A multi-institutional experience with stereotactic radiosurgery for solitary brain metastasis. Int J Radiat Oncol Biol Phys 1994;28:797–802.

5   Flickinger JC, Kondziolka D, Maitz AH, Lunsford LD: Analysis of neurological sequelae from radiosurgery of arteriovenous malformations: How location effects outcome. Int J Radiat Oncol Biol Phys 1997;8 (in press).

6   Flickinger JC, Kondziolka D, Pollock BE, Lunsford LD: Evolution of technique for vestibular schwannoma radiosurgery and effect on outcome. Int J Radiat Oncol Biol Phys 1996;36:275–280.

7   Flickinger JC, Kondziolka D, Pollock BE, Maitz AH, Lunsford LD: Complications from arteriovenous malformation radiosurgery: Multivariate analysis and risk modeling. Int J Radiat Oncol Biol Phys 1997;38:485–490.

8   Flickinger JC, Lunsford LD, Kondziolka D, Maitz AM, Epstein A, Simons S, Wu A: Radiosurgery and brain tolerance: An analysis of neurodiagnostic imaging changes following gamma knife radiosurgery for arteriovenous malformations. Int J Radiat Oncol Biol Phys 1992;23:19–26.

9   Flickinger JC, Lunsford LD, Kondziolka D: Radiosurgical dosimetry: Principles and clinical implications; in DeSalles AF, Goetsch S (eds): Stereotactic Surgery and Radiosurgery. Madison, Medical Physics Publishing, 1993, pp 293–306.

10 Flickinger JC, Lunsford LD, Linskey ME, Duma CM, Kondziolka D: Gamma knife radiosurgery for acoustic tumors: Multivariate analysis of four-year results. Radiother Oncol 1993;27:91–98.

11 Flickinger JC, Pollock BE, Kondziolka D, Lunsford LD: A dose-response analysis of arteriovenous malformation obliteration after radiosurgery. Int J Radiat Oncol Biol Phys 1996;36:873–879.

12 Flickinger JC, Schell MC, Larson DA: Estimation of complications for linear accelerator radiosurgery with the integrated logistic formula. Int J Radiat Oncol Biol Phys 1990;19:143–148.

13 Hellman S: Principles of cancer management: Radiation therapy; in DeVita VT, Hellman S, Rosenberg SA (eds): Cancer: Principles and Practice of Oncology, ed 5. Philadelphia, Lippincott-Raven, 1997, p 320.

14 Kjellberg R, Hanamura T, Davis K, Lyons S, Butler W, Adams R: Bragg-peak proton-beam therapy for arteriovenous malformations of the brain. N Engl J Med 1983;309:269.

15 Linskey ME, Flickinger JC, Lunsford LD: The relationship of cranial nerve length to the development of delayed facial and trigeminal neuropathies after stereotactic radiosurgery for acoustic tumors. Int J Radiat Oncol Biol Phys 1993;15:227–234.

16 Loeffler JS, Alexander E, Siddon R, Saunders W, Coleman N, Winston K: Stereotactic radiosurgery for intracranial arteriovenous malformations using a standard linear accelerator. Int J Radiat Oncol Biol Phys 1989;17:673–677.

17 Loeffler JS, Shrieve DC: What is appropriate therapy for a patient with a single brain metastasis? Int J Radiat Oncol Biol Phys 1994;29:915–917.

18 Pollock BE, Flickinger JC, Lunsford LD, Kondziolka D: Hemorrhage risk after radiosurgery for arteriovenous malformations. Neurosurgery 1996;38:652–661.

19 Shaw E, Scott C, Souhami L, et al: Radiosurgery for the treatment of previously irradiated recurrent primary brain tumors and brain metastases: Initial report of Radiation Therapy Oncology Group protocol (90-05). Int J Radiat Oncol Biol Phys 1996;34:647–654.

20 Tishler RB, Loeffler JS, Lunsford LD, Duma C, Alexander E III, Kooy HM, Flickinger JC: Tolerance of cranial nerves of the cavernous sinus to radiosurgery. Int J Radiat Oncol Biol Phys 1993;27:215–221.

John C. Flickinger, MD, Department of Radiation Oncology,
200 Lothrop Street, Pittsburgh, PA 15213 (USA)
Tel. (412) 647 3600, Fax (412) 647 6029

Lunsford LD, Kondziolka D, Flickinger JC (eds): Gamma Knife Brain Surgery.
Prog Neurol Surg. Basel, Karger, 1998, vol 14, pp 51–59

....

Patient Outcomes after Arteriovenous Malformation Radiosurgery

Bruce E. Pollock

Department of Neurologic Surgery, Mayo Clinic, Rochester, Minn., USA

The primary goal in the management of arteriovenous malformations (AVMs) is to eliminate the risk of hemorrhage for the patient. Studies on the natural history of untreated AVMs have demonstrated a 2–4% annual hemorrhage rate, and a mortality rate of approximately 30% per bleed [1, 19]. Factors reportedly associated with an increased bleeding risk include history of a prior hemorrhage, venous outflow restriction, and AVM morphology [17, 20]. Surgical resection of AVMs is the preferred management strategy for patients with AVMs that can be resected with a low morbidity. The greatest advantage of surgical resection is that the bleeding risk for the patient is eliminated the day of surgery if the AVM is completely removed. However, the number of patients having incomplete AVM resection is rarely reported in most surgical series [9, 28]. Patients undergoing microsurgical resection are at risk for the medical complications associated with intracranial surgery (infection, seizures, pneumonia, deep vein thrombosis).

Recognizing the difficulties associated with neurological surgery in the 1950s, Dr. Lars Leksell sought a less invasive method to treat neurologic diseases. His development and refinements of the initial radiosurgical devices led to the development of the Gamma Knife®. One of the first indications for radiosurgery was AVMs. This was because AVMs could be visualized and stereotactically targeted with cerebral angiography; the first radiosurgical cases predated computed tomography (CT) and magnetic resonance imaging (MRI) by more than a decade. The initial radiation doses chosen for AVM radiosurgery were not based on any available knowledge on the radiobiology of high-dose, single-fraction radiation. Instead, an earlier report on radiotherapy in the treatment of tumors of the neck found that 60 Gy delivered in fractionated doses was associated with vessel rupture and hemorrhage [11]. To avoid such complications, a maximum dose of 50 Gy was chosen. This initial dose selection is quite similar to the maximum radiation dose utilized in contemporary AVM radiosurgery.

This chapter will outline the pertinent factors involved in AVM radiosurgery, compare patient outcomes after surgical resection to radiosurgery, and outline the current expectations of AVM radiosurgery.

AVM Obliteration after Radiosurgery

Radiosurgery damages the endothelial cells within the AVM nidus [27]. In the months and years that follow, the irradiated vessels develop a concentric narrowing of their lumina by fibrous proliferation, foam cell accumulation, and hyalinization that results in vessel occlusion. Clinical reports have provided a number of basic principles regarding dose selection for AVM radiosurgery [2, 7, 11, 16, 29, 30]. First, there is a direct correlation between the minimum dose to the AVM and the rate of complete obliteration. Second, the higher the minimum dose, the more rapid AVM obliteration occurs. Third, minimum doses above 25 Gy are unlikely to improve obliteration rates and are associated with a higher rate of radiation-related complications. Fourth, larger AVMs have lower obliteration rates in most radiosurgical series because they received lower radiation doses, not because of their innate radiobiologic resistance. Recognition of the dose-volume relationship in the development of postradiosurgical radiation-related complications has led most centers to base their dose selection for AVM radiosurgery on predictions for radiation necrosis by either Kjellberg's 1% iso-effect line [12] or Flickinger's integrated logistic formula [3].

For AVM obliteration to occur after radiosurgery, the radiation dose must not only be of a sufficient amount, but it must also be well targeted to the nidus. In an analysis of 45 patients undergoing repeat AVM radiosurgery, Pollock et al. [24] found that over half of the patients had AVM nidus outside the original irradiated volume. Reliance on conventional two-dimensional angiography as the sole imaging modality for AVM radiosurgery can result in the overestimation of the nidus volume if adjacent draining veins or normal brain parenchyma are included in the dose plan. It has been shown that the addition of either stereotactic CT or MRI to the neuroimaging database is critical to better define the actual three-dimensional shape of most AVMs [13]. Flickinger et al. [5] analyzed 197 AVM patients to better define the dose response of AVMs after radiosurgery. They found that when the accuracy of AVM targeting was considered, complete obliteration occurred between 70 and 90% when the AVM marginal dose was from 16 to 20 Gy.

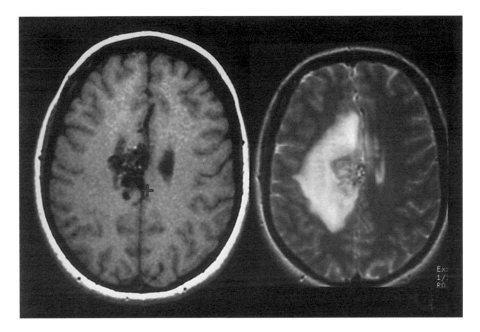

Fig. 1. Axial MRIs of a 30-year-old woman with an AVM located in the corpus callosum and right medial frontal lobe. Left: T1-weighted MRI at the time of radiosurgery. A radiation dose of 18 Gy was delivered to the AVM margin (11.7 cm³). Right: T2-weighted MRI 12 months after radiosurgery. Note the area of increased signal adjacent to the irradiated AVM. The patient developed left leg weakness and headaches and required a course of corticosteroids.

Postradiosurgical Radiation-Related Complications

The likelihood of complete AVM obliteration must be balanced against the need to avoid an unacceptable rate of radiation-related complications after radiosurgery. Flickinger et al. [4] found that 30% of AVM patients developed new areas of increased signal on long-TR images adjacent to or within the irradiated volume at a median of 8 months after radiosurgery. Overall, 10% of these patients developed symptoms (new neurologic deficit, headache) which related to the imaging changes (fig. 1). Approximately half of the symptoms were transient and resolved completely for a permanent radiation-related complication rate of 5%. Multivariate analysis of factors found that the volume of tissue that received at least 12 Gy (the 12-Gy volume) and AVM location (thalamus and brainstem being worse than cerebral or cerebellar hemispheres) were associated with the development of symptomatic radiation-related complications. Lax and Karlsson [15] also found a relationship between the average radiation dose to the 20 cm^3 centered around the target and the development of radiation-related complications. Radiation-related complications occurred in 13 of 367 peripherally located AVMs (4%) compared to 32 of 456 centrally located AVMs (7%). Although it has not been shown scientifically that conformal dose planning to AVMs will result in lower complication rates, the exclusion of the surrounding normal brain from the dose plan is theoretically advantageous. Another confounding factor is that many of the imaging changes we observe after radiosurgery may be due to alterations in perilesional blood flow as the nidus progresses to obliteration instead of radiation damage to the region. In such cases, it is possible that improved dose planning would have little effect on the development of the imaging changes after radiosurgery.

Postradiosurgical AVM Hemorrhage

The primary drawback of radiosurgery as a management strategy for AVMs is that the patient remains at risk for bleeding until the nidus has obliterated. The latency interval after radiosurgery obliteration until obliteration occurs generally ranges from 1 to 3 years, although AVM obliteration has been documented as early as 6 months and as late as 4 years. Earlier reports on AVM radiosurgery suggested that the annual hemorrhage rate after radiosurgery prior to obliteration was greater than the natural history of untreated AVMs [2, 12, 25]. Recently, three separate reports have analyzed postradiosurgical AVM hemorrhage rates [6, 10, 21]. Pollock et al. [21] and Friedman et al. [6] both found that radiosurgery did not change the annual bleeding rates during the latency interval. Patients with a history of a prior bleed were no more likely to sustain a postradiosurgical hemorrhage than patients who presented with seizures or headaches. Of note, Friedman et al. [6] pointed out the important fact that it would require over 700 patient-years of follow-up to detect a reduction in the annual bleeding rate from 4 to 3% at a significance level of 0.05. Karlsson et al. [10] reported 1,604 patients with a total of 2,340 patient-years at risk for postradiosurgical AVM hemorrhage. They detected a decrease in the AVM hemorrhage rates within 6 months after radiosurgery, and patients who received higher radiation doses were conferred the greatest protection from bleeding. Prior hemorrhage was not found to be a significant predictor of postradiosurgical bleeding. When considered together, these three reports with 2,000 patients and more than 3,000 patient-years at risk for hemorrhage document quite convincingly that radiosurgery does not increase the annual bleeding rate for AVM patients prior to obliteration. In fact, the data suggests that even patients with incomplete AVM obliteration after radiosurgery receive some protection against the risk of future bleeding.

Comparison of AVM Microsurgery and Radiosurgery

It is difficult to directly compare the results of microsurgical and radiosurgical series on AVMs. Grading scales designed to predict patient outcomes after surgical resection of AVMs can be applied to radiosurgical series, but they are insensitive to important factors associated with successful AVM radiosurgery [28]. Also, patients undergoing radiosurgery are frequently considered to be poor candidates for surgical resection because their AVMs are in critical brain locations. As a result, a selection bias is present in most seers with 'good' AVMs undergoing surgical resection, whereas 'bad' AVMs are referred for radiosurgery. Morgan et al. [18] analyzed 92 consecutive AVM patients undergoing surgical resection between 1989 and 1996. They found that patients with Spetzler-Martin (SM) Grade III AVMs with lenticulostriate arterial supply had a >80% complication rate, compared to a 15% complication for patients with SM Grade III AVMs without this blood supply. Likewise, Lawton et al. [14] reported 32 patients with AVMs located in the thalamus, basal ganglia, or brainstem. For patients managed before the availability of radiosurgery, the complete resection rate was 43% (4 of 9 patients). The AVM cure rate after the availability of radiosurgery increased to 72% (13 of 18 patients), with the follow-up on the 5 remaining patients being too short for postradiosurgical angiography. Conversely, Pollock et al. [25] reported 65 patients managed with radiosurgery for SM Grade I-II AVMs. Of 32 patients, 31 (97%) evaluable by follow-up angiography had complete nidus obliteration (4 patients had an early draining vein), and no patient sustained a radiation-related complication. Five patients had a postradiosurgical hemorrhage: 2 patients died, 3 recovered without new neurologic deficit. Radiosurgery may also be less likely to result in neurologic deficits for patients with AVMs located in critical cortical structures such as sensorimotor or visual cortex. Of 34 patients, only 2 (6%) with AVMs in the postgeniculate visual pathways developed new visual field defects after radiosurgery [26].

Expectations of Contemporary AVM Radiosurgery

The ideal patient outcome after AVM radiosurgery is complete obliteration without new neurologic deficits. Studies on AVM radiosurgery traditionally have reported the angiographic obliteration rate, the incidence of radiation-related complications, and the number of postradiosurgical AVM hemorrhages [2, 7, 16, 29, 30]. It is impossible to determine from these studies the number of patients with complications related to radiosurgery (radiation-related or postradiosurgical hemorrhage) in relation to the obliteration status of their

Table 1. Patient outcomes after AVM radiosurgery

Outcome	Definition
Excellent	Complete AVM obliteration, no new neurologic deficit
Good	Complete AVM obliteration, new minor neurologic deficit that does not interfere with the patient's preoperative level of functioning
Unchanged	Incomplete AVM obliteration, no new neurologic deficit
Poor	Incomplete AVM obliteration, new minor neurologic deficit that does not interfere with the patient's preoperative level of functioning
Disabled	New major neurologic deficit that interferes with the patient's preoperative level of functioning, regardless of the status of the AVM
Dead	

AVMs. Table 1 outlines standardized patient outcomes after AVM radiosurgery. Patient outcomes can be predicted by multiplying the probabilities of AVM obliteration, radiation-related complications, and postradiosurgical bleeding (table 2). Pollock et al. [23] have shown that for patients with AVMs measuring <2.0 cm in average diameter (volume <4 cm^3) located in the cerebral or cerebellar hemispheres, 80% of patients will achieve an excellent outcome after a single radiosurgical procedure.

Patients with large volume AVMs (>3 cm in average diameter) should be strongly considered for surgical resection if the risk is felt to be acceptable. Patients with large AVMs in critical brain locations that are poor candidates for surgical removal can be managed by several different radiosurgical strategies. First, the AVM can be covered with a low marginal dose (>15 Gy) at the initial radiosurgical procedure accepting that the complete obliteration rate will also be low (approx. 50%). Repeat radiosurgery can then be performed 3–4 years later to the residual nidus that tends to be much smaller. The obliteration rate for repeat AVM radiosurgery is 60–70%, for an overall obliteration rate of 70–80%. Second, the patient can undergo preradiosurgical embolization to reduce the size of the AVM nidus to improve the results of radiosurgery. However, for difficult AVMs the complication rate of embolization can be quite high, and re-canalization of the embolized nidus can occur in 10–20% of patients [8]. Third, large AVMs can be volume-staged with separate components of the AVM being irradiated in multiple procedures over

Table 2. Prediction of successful AVM radiosurgery (A) and expected excellent outcomes after single-session AVM radiosurgery utilizing contemporary dose planning and dose prescription (B)

A. Prediction of successful AVM radiosurgery
  (Probability of AVM obliteration) ×
  (Probability of no deficit from postradiosurgical AVM hemorrhage) ×
  (Probability of no deficit from radiation-related changes).

B. Expected outcomes after single-session AVM malformation
(1) 1 cm average diameter AVM not in the brainstem/thalamus:
     $(0.95) \times (0.97) \times (0.99) = 0.91$
(2) 2 cm average diameter AVM not in the brainstem/thalamus:
     $(0.85) \times (0.97) \times (0.98) = 0.81$
(3) 3 cm average diameter AVM not in the brainstem/thalamus:
     $(0.60) \times (0.93) \times (0.94) = 0.52$
(4) 1 cm average diameter AVM in the brainstem/thalamus:
     $(0.95) \times (0.97) \times (0.96) = 0.88$
(5) 2 cm average diameter AVM in the brainstem/thalamus:
     $(0.85) \times (0.97) \times (0.92) = 0.76$

---

a 3- to 6-month interval. This strategy will hopefully result in a higher overall obliteration rate with acceptable radiation-related complications. Nonetheless, any intervention for such complex AVMs should only be contemplated if the patient has sustained a prior hemorrhage, or if the patient has a progressive neurologic decline related to the AVM.

The Pittsburgh Arteriovenous Malformation Radiosurgery (PAR) grading scale was developed to predict patient outcomes after AVM radiosurgery [22]. Based on the coefficients of five significant factors (patient age, AVM volume, AVM location, number of draining veins, prior embolization) from a multivariate analysis of patient outcomes after AVM radiosurgery, the PAR grading scale was found to significantly correlate with patient outcomes. This grading scale may prove useful at predicting patient outcomes after AVM radiosurgery, but at present has two primary drawbacks: (1) the scale must be validated by other radiosurgical centers, and (2) the PAR grading scale in its present configuration is cumbersome and difficult to use. Karlsson et al. [11] have advanced the K constant as a means to predict AVM obliteration after radiosurgery. The drawbacks of the K constant are also twofold: (1) because it is based on factors determined at the time of radiosurgery (the marginal AVM radiation dose), it is not useful in discussing management options with patients prior to surgery, and (2) the K constant does not account for either radiation-related complications or postradiosurgical hemorrhage and their impact on patient outcomes after radiosurgery. Thus, although we have learned a great deal about AVM radiosurgery in the past 20 years, much work remains to be done to develop a simple, statistically sound method to predict patient outcomes after AVM radiosurgery.

Conclusions

Radiosurgery has been proven to be a safe and effective management for many AVM patients. Over the past decade, many of the myths concerning AVM radiosurgery have been dismissed and a great deal of information is now available on obliteration, radiation-related complications, and postradiosurgical hemorrhage. Future studies on AVM radiosurgery should report patient outcomes by a method that incorporates all the factors involved in successful AVM radiosurgery.

References

1   Brown RD, Wiebers DO, Forbes G, O'Fallon WM, Piepgras DG, Marsh WR, Maciunias RJ: The natural history of unruptured intracranial arteriovenous malformations. J Neurosurg 1988;68: 352–357.
2   Colombo F, Pozza F, Chierego G, Casentini L, De Luca G, Francescon P: Linear accelerator radiosurgery of cerebral arteriovenous malformations: An update. Neurosurgery 1994;34:14–21.
3   Flickinger JC: An integrated logistic formula and prediction of complications from radiosurgery. Int J Radiat Oncol Biol Phys 1989;17:879–885.
4   Flickinger JC, Kondziolka D, Pollock BE, Maitz A, Lunsford LD: Complications from arteriovenous malformation radiosurgery: Multivariate analysis and risk modeling. Int J Radiat Oncol Biol Phys 1997;38:485–490.
5   Flickinger JC, Pollock BE, Kondziolka D, Lunsford LD: A dose-response analysis of arteriovenous malformation obliteration after radiosurgery. Int J Radiat Oncol Biol Phys 1996;36:873–879.
6   Friedman WA, Blatt DL, Bova FJ, Buatti JM, Mendenhall WM, Kublis PS: The risk of hemorrhage after radiosurgery for arteriovenous malformations. J Neurosurg 1996;84:912–919.
7   Friedman WA, Bova FJ, Mendenhall WM: Linear accelerator radiosurgery for arteriovenous malformations: The relationship of size to outcome. J Neurosurg 1995;82:180–189.
8   Gobin YP, Laurent A, Merienne L, Schlienger M, Aymard A, Houdart E, Casasco A, Lefkopoulos D, George B, Merland JJ: Treatment of brain arteriovenous malformations by embolization and radiosurgery. J Neurosurg 1996;85:19–28.
9   Hamilton MG, Spetzler RF: The prospective application of a grading system for arteriovenous malformations. Neurosurgery 1994;34:2–7.
10 Karlsson B, Lindquist C, Steiner L: The effect of gamma knife surgery on the risk of rupture prior to AVM obliteration. Minim Invasive Neurosurg 1996;39:21–27.
11 Karlsson B, Lindquist C, Steiner L: Prediction of obliteration after gamma knife surgery for cerebral arteriovenous malformations. Neurosurgery 1997;40:425–431.
12 Kjellberg RN, Hanamura T, Davis KR, Lyons SL, Adams RD: Bragg-peak proton-beam therapy for arteriovenous malformations of the brain. N Engl J Med 1983;309:269–274.
13 Kondziolka D, Lunsford LD, Kanal E, Talagala L: Stereotactic magnetic resonance angiography for targeting in arteriovenous malformation radiosurgery. Neurosurgery 1994;35:585–591.

14   Lawton MT, Hamiliton MG, Spetzler RF: Multimodality treatment of deep arteriovenous malformations: Thalamus, basal ganglia, and brain stem. Neurosurgery 1995;37:29–36.

15   Lax I, Karlsson B: Prediction of complications in gamma knife radiosurgery of arteriovenous malformations. Acta Oncol 1996;35:49–56.

16   Lunsford LD, Kondziolka D, Flickinger JC, Bissonette DJ, Jungreis CA, Vincent D, Pentheny S, Horton JA: Stereotactic radiosurgery for arteriovenous malformations of the brain. J Neurosurg 1991;75:512–524.

17   Miyasaka Y, Yada K, Ohwada T, Kithara T, Kurata A, Irikura K: An analysis of the venous drainage system as a factor in the hemorrhage from arteriovenous malformations. J Neurosurg 1992;76:239–243.

18   Morgan MK, Drummond KJ, Grinnell V, Sorby W: Surgery for cerebral arteriovenous malformation: Risks related to lenticulostriate arterial supply. J Neurosurg 1997;86:801–805.

19   Ondra SL, Troupp H, George ED, Schwab K: The natural history of symptomatic arteriovenous malformations of the brain: A 24-year follow-up assessment. J Neurosurg 1990;73:387–391.

20   Pollock BE, Flickinger JC, Lunsford LD, Bissonette DJ, Kondziolka D: Factors that predict the bleeding risk of cerebral arteriovenous malformations. Stroke 1996;27:1–6.

21   Pollock BE, Flickinger JC, Lunsford LD, Bissonette DJ, Kondziolka D: Hemorrhage risk after stereotactic radiosurgery of cerebral arteriovenous malformations. Neurosurgery 1996;38:652–661.

22   Pollock BE, Flickinger JC, Lunsford LD, Maitz A, Kondziolka D: The Pittsburgh Arteriovenous Malformation Radiosurgery (PAR) grading scale; in Kondziolka D (ed): Radiosurgery 1997. Radiosurgery. Basel, Karger, 1998, vol 2, pp 137–146.

23   Pollock BE, Flickinger JC, Lunsford LD, Maitz AH, Kondziolka D: Factors associated with successful arteriovenous malformation radiosurgery. Neurosurgery 1997;8 (in press).

24   Pollock BE, Kondziolka D, Lunsford LD, Bissonette DJ, Flickinger JC: Repeat stereotactic radiosurgery of arteriovenous malformations: Factors associated with incomplete obliteration. Neurosurgery 1996;38:318–324.

25   Pollock BE, Lunsford LD, Kondziolka D, Maitz AH, Flickinger JC: Patient outcomes after stereotactic radiosurgery for 'operable' arteriovenous malformations. Neurosurgery 1994;35:1–8.

26   Pollock BE, Lunsford LD, Kondziolka D, Bissonette DJ, Flickinger JC: Stereotactic radiosurgery for postgeniculate visual pathway arteriovenous malformations. J Neurosurg 1996;84:437–441.

27   Schneider BF, Eberhard DA, Steiner LE: Histopathology of arteriovenous malformations after gamma knife radiosurgery. J Neurosurg 1997;87:352–357.

28   Spetzler RF, Martin NA: A proposed grading system for arteriovenous malformations. J Neurosurg 1986;65:476–483.

29   Steiner L, Lindquist C, Adler JR, Torner JC, Alves W, Steiner M: Clinical outcome of radiosurgery for cerebral arteriovenous malformations. J Neurosurg 1992;77:1–8.

30   Yamamoto Y, Coffey RJ, Nichols DA, Shaw EG: Interim report on the radiosurgical treatment of cerebral arteriovenous malformations. The influence of size, dose, time and technical factors on obliteration rate. J Neurosurg 1995;83:832-837.

Bruce E. Pollock, MD, Department of Neurologic Surgery, Mayo Clinic,
200 First Street, SW, Rochester, MN 55905 (USA)
Tel. (507) 284 2775, Fax (507) 284 5206

Lunsford LD, Kondziolka D, Flickinger JC (eds): Gamma Knife Brain Surgery.
Prog Neurol Surg. Basel, Karger, 1998, vol 14, pp 60–77

....

Gamma Knife Radiosurgery and Particulate Embolization of Dural Arteriovenous Fistulas, with a Special Emphasis on the Cavernous Sinus

Robert J. Coffey[a], *Michael J. Link*[c], *Douglas A. Nichols*[b], *Bruce E. Pollock*[a], *Deborah A. Gorman*[a]

Department of [a]Neurologic Surgery and [b]Diagnostic Radiology, Mayo Clinic and Mayo Foundation, Rochester, Minn., and [c]Department of Neurologic Surgery, Mayo Clinic, Jacksonville, Fla., USA

For nearly 30 years, neurosurgeons and neuroradiologists have recognized dural arteriovenous fistulas (DAVFs) to be a distinct class of acquired vascular malformations that affect the dural venous sinuses and their tributaries [32, 40, 63, 80]. Over the same period, observational, clinical, radiographic, anatomic, and more recently, laboratory studies have revealed much information regarding the pathogenesis, natural history, and appropriate management of this disorder. In the 1960s and 1970s, astute clinicians observed the participation of meningeal arteries and dural veins in certain intracranial arteriovenous malformations [63]. More importantly, Houser et al. [31, 32] noted the de novo appearance of DAVFs in adult patients who had undergone previous carotid angiograms (for other reasons) that did not show the lesion beforehand. Almost invariably, the transverse-sigmoid DAVFs that they described were associated with complete or partial thrombosis of the involved sinus. Thus arose the current model of venous sinus thrombosis as the initiating event in fistula formation. Studies of the dural microvasculature, and recent animal studies involving the creation of venous hypertension have elucidated the possible anatomic substrate and physiological mechanisms involved in fistula formation as well [8, 14, 25, 29, 36, 37, 40, 47, 61, 75, 80]. These include venous hypertension as the stimulus for the release of chemical angiogenesis factors which in turn cause dilation or formation of dural venous collaterals that drain into the remaining unobstructed segments of the venous sinuses.

Table 1. Reported location of DAVFs (review of 377 cases)	Location	Percent of total
	Transverse – sigmoid	62.6
	Cavernous sinus	11.9
	Tentorium – incisura	8.4
	Convexity – superior sagittal sinus	7.4
	Anterior fossa	5.8
	Sylvian – middle fossa	3.7

Table 2. Reported frequency of hemorrhage for DAVFs	Location	Percent presenting with hemorrhage
	Tentorium – incisura	80–90
	Anterior fossa	60–80
	Sylvian – middle fossa	60–70
	Convexity – superior sagittal sinus	50–60
	Transverse – sigmoid	15–20
	Cavernous sinus	10–15

The inciting events or conditions that cause dural venous sinus thrombosis and/or induce fistula formation are diverse and include local or regional infection and/or inflammatory processes (otitis, mastoiditis), trauma (convexity or basal skull fractures, minor closed head injury), or iatrogenic causes (scalp surgery, twist drill penetration of the skull). Trauma and surgical intervention represent special circumstances under which small direct fistulas may arise rather quickly – setting in motion the cascade of events that cause progressive fistula growth as in the spontaneous cases [34, 85]. Recent reviews have tabulated the location, presenting symptoms, and most importantly, the risk of hemorrhage associated with DAVFs (tables 1, 2). Of course, even meta-analysis of clinical, interventional neuroradiologic, or surgical series inevitably introduce both an ascertainment bias and a selection bias into the current discussion. These limitations notwithstanding, lesions of the transverse and sigmoid sinuses are the most commonly diagnosed DAVFs. Patients present with pulsatile and/or intractable tinnitus or bruit, and unless an unusual degree of retrograde pial/cortical venous drainage is present, have a relatively low risk of hemorrhage. Nevertheless, the noise in these patients' head often is so unbearable that curative (or at least palliative) treatment is worthwhile. Carotid cavernous DAVFs that receive arterial supply from branches of the external carotid artery (and/or dural branches of the cavernous internal carotid artery) were once included among the 'spontaneous' carotid cavernous fistulas in older references. These occur less frequently than transverse-sigmoid DAVFs and also have a low risk of hemorrhage unless cortical/pial venous drainage is present. Patients with cavernous sinus region DAVFs may develop an alarming symptom complex consisting of proptosis, chemosis, extraocular motion palsies, and sometimes, threatened visual loss from increased intraocular pressure. In these patients, curative treatment is indicated for visual salvage in addition to cosmetic and other symptomatic considerations. In contrast, DAVFs that involve the veins of the anterior cranial fossa, and those located along the tentorial incisura (which often have cortical or pial venous drainage) are believed by some authors to remain clinically silent until they cause an intracranial subarachnoid hemorrhage.

Treatment and Rationale

The variety of treatments for DAVFs has included observation alone (patient reassurance with no treatment) [1, 6, 15, 27, 38, 43, 45, 53, 64, 81, 84], intermittent manual compression of the carotid artery or scalp arteries [2, 5, 20, 26, 33, 36, 50, 55, 59, 72, 76], transarterial or transverse embolization [7, 12, 13, 17–19, 21–24, 28, 35, 39, 43, 45, 54, 56, 57, 60, 62, 65, 66, 68–70, 74, 77, 79, 81–83], surgical resection or venous disconnection, and more recently, stereotactic radiosurgery [3, 10, 16, 30, 42, 48, 49, 58, 71]. Immediate or early curative treatment has been advocated for patients who experience a hemorrhage, for patients who have prominent cortical or pial venous drainage (incipient or threatened hemorrhage) and for patients who have rapidly progressive neurological signs or symptoms [2, 41, 42, 54]. Owing to historical factors including clinical interest, a long series of publications, and physician referral patterns, the Mayo Clinic has accumulated a large experience in the evaluation and treatment of patients with DAVFs [9, 31, 32, 49, 55, 72, 78]. Cerebrovascular surgeons at our institution have reported excellent results, especially for selected high-risk patients with transverse-sigmoid lesions. However, because conventional surgery sometimes was associated with major morbidity [72] and because lesions in some locations (especially the cavernous and petrosal sinuses) were not easily accessible to effective surgical management, we developed other treatment strategies for selected patients with an unfavorable surgical risk/benefit ratio. This experience led to the development of a management strategy that has included radiosurgery and endovascular occlusion using a novel staging approach developed at the Mayo Clinic.

While successful endovascular approaches from either or both the arterial and venous side of DAVFs have been reported, long-term follow-up in such patients is just now becoming available [7, 12, 13, 21, 22, 28, 35, 43, 54, 60, 62, 65, 66, 68, 69, 73, 76, 79, 81, 83]. In some studies, late recanalization of embolized DAVFs has been documented [4, 18, 19, 22, 28, 32, 41, 43, 58, 67, 77, 81–83]. After a few sporadic reports of small series at other institutions described variable results after LINAC radiosurgery of DAVFs, we reported our experience using gamma knife radiosurgery (usually combined with particulate arterial embolization) in 29 patients, the largest series to date [49]. Our approach included (a) radiosurgery alone for more indolent lesions without pial venous drainage or for those lesions without an external carotid artery supply accessible to embolization, and (b) radiosurgery 1–2 days before particulate embolization in patients with pial venous drainage or in patients with symptoms that required early palliation. The rationale for this approach was that radiosurgery alone was expected to cause obliteration of most DAVFs between 1 and 3 years after treatment, and secondly, that radiosurgery before embolization allowed optimal angiographic demonstration of the lesion for dose planning, simultaneously avoiding the pitfall of having embolization temporarily obscure portions of the nidus that otherwise might have gone untreated by the radiosurgical procedure. Patients with pial venous drainage underwent particulate embolization after radiosurgery to provide early protection against hemorrhage during the latency period before radiation-induced obliteration. Even if recanalization of the embolized portions were to occur, the fistula would be undergoing radiosurgery-induced obliteration simultaneously. In patients with intractable bruit, orbital venous congestion and/or diplopia, embolization provided palliation by reducing disabling symptoms, often permanently.

Materials and Methods

Patient Population

Between October 1990 and September 1997 we treated 67 patients with intracranial DAVFs using the Leksell Gamma Unit (Model U, 1990–January 1997; Model B, after March 1997) at the Mayo Clinic in Rochester, Minnesota. Our patients included 24 men and 43 women. The mean patient age was 61 years (range 20–80 years). All patients were symptomatic from their fistulas. Their clinical presentation and locations of the fistulas are summarized in tables 3 and 4. Comparison of our series with table 1 reveals that cavernous sinus lesions were overrepresented in our case material. The reasons for this will become obvious later in this chapter. Before referral for radiosurgery, 5 patients underwent a variety of noncurative procedures that included surgery alone (1 patient), embolization alone (2 patients) and embolization plus surgery (2 patients). Those patients whose initial presentation included hemorrhage had chronic neurologic deficits, and were in stable clinical condition at the time of treatment.

Table 3. Clinical features of 67 patients with DAVFs treated using Gamma knife radiosurgery[1]

Symptoms/signs	Patients, n
Intractable bruit or vascular headache	43
Chemosis, ophthalmoplegia or visual loss	26
Hemorrhage	7
Altered mental status	3
Seizures	2
Other neurologic deficit	3

[1] Some patients experienced more than one category of symptoms.

Table 4. Locations of 67 DAVFs treated using Gamma knife radiotherapy

Location	Patients	
	n	%
Transverse – sigmoid	24	36
Cavernous sinus	20	30
Jugular bulb	7	10
Petrosal sinus	7	10
Torcular – superior sagittal sinus	5	7
Tentorial notch	2	3
Cranial convexity	2	3

Within several days (usually 2 or less) after radiosurgery, 40 patients underwent particulate embolizaion of accessible external carotid artery feeding branches for the indications listed in table 5.

Patient Evaluation

All patients underwent a thorough neurologic examination and pertinent specialty examinations by an ophthalmologist and/or otolaryngologist. Objective tests of visual acuity, visual fields, pure-tone audiometry, speech discrimination, and other functional assessments were carried out as indicated to establish baseline data for comparison during follow-up.

Owing to the complexity of the DAVF cases referred to our institutions, and owing to occasionally erroneous diagnoses assigned to them beforehand, an intensive review of available imaging studies always was performed. In many cases repeat angiography was necessary before a decision could be made regarding treatment. Bilateral selective internal carotid, external carotid, and vertebral angiographic series were essential for diagnosis *before* proceeding with headframe application and radiosurgery. In a few instances, superselective branch-

Table 5. Reasons for
embolization in 40 patients
with DAVFs treated using
Gamma knife radiosurgery[1]

Symptoms/signs/findings	Patients, n
Retrograde cortical or pial venous drainage	20
Intractable bruit	26
Chemosis, orbital congestion, visual symptoms	21

[1] Most patients had more than one category of symptoms, signs, or findings.

vessel angiograms also were essential to sort out the exact site of a fistula, and/or its appropriateness for radiosurgery or embolization. Thus, the final decision to proceed with our treatment program was made only after a thorough analysis of the clinical and imaging features of each case on an individual basis.

Leksell Gamma Knife Dose Planning and Evolution of Technique

Conventional cut film biplane angiography using high-resolution magnification and photographic subtraction techniques was employed in all cases. Selective internal carotid, external carotid, and/or vertebral artery injections were performed as required to demonstrate the target for dose planning. Superselective angiography of external carotid artery branches was also performed when necessary in unusual cases. During our early experience, the KULA dose-planning system produced scaled isodose overlay transparencies for super-imposition upon the angiogram films. Beginning in early 1993, the Leksell Gamma Plan System® (with successive software versions and updates) allowed transmission scanning of the angiogram films directly into the dose-planning computer. Stereotactic MRI (or CT when MRI was contraindicated) with direct network or DAT tape data transfer to the dose-planning computer now is routine for all vascular cases. In our early experience, these essential aids to dose planning were reserved for instances where the fistula nidus was very complex, or located near cranial nerves or the brainstem. At present, the ability to correlate angiographic and MR imaging with the targets and dose plan almost instantaneously makes the radiosurgical planning of these unusual lesions more conformal and possibly even more safe and effective than beforehand. Because the cross-sectional profile of significant portions of the dural venous sinus system is triangular (for example the superior sagittal sinus, transverse sinus, sigmoid sinus, and even the cavernous sinus) and because this triangular profile follows the interior curvature of the cranial vault and skull base, the dose plans for large or extensive DAVFs can appear extraordinarily complex at first glance. In some cases, neither high-resolution angiography nor MRI can show for certain whether the actual fistula sites are located on an external dural sinus surface facing the skull or an interior surface facing the subarachnoid space. In fact, contributions from meningeal arteries or transosseous scalp or skull base arteries can involve any/or all three surfaces of a venous sinus. Thus, an adequate dose plan sometimes must cover the entire cross-sectional profile of a substantial length of sinus, yielding the appearance of a twisted or curved sausage, especially for transverse-sigmoid-jugular fistulas. Under these circumstances, a curious phenomenon occurs: the target receives only the margin dose, while the highest central radiation dose is administered to the blood elements flowing through the sinus. Another reason for the apparent safety of radiosurgery for even large DAVFs accrues from a corollary to this phenomenon: most fistulas are surrounded by nonneurologic tissue (bone, blood, cerebrospinal fluid, scalp) for 180–270° of the dose-plan circumference.

The mean radiation dose to the margin of the angiographically defined fistulas was 19 Gy (range 15–20 Gy) at the 50% or higher isodose line using 1–11 isocenters (mean 4). In order to reach target coordinates for the distal sigmoid sinus and jugular bulb in the Model U Gamma knife, we often had to apply the Leksell Model G headframe with the anterior pins in the maxillary prominences. This avoided otherwise inevitable collisions between the anterior posts and the Model U collimator helmets. Such collisions are less of an issue with the newer Model B Gamma knife, although proper headframe placement remains essential.

Embolization and Follow-Up Plan

Particulate embolization was used as a planned adjunct within a few days after radiosurgery in 40 patients in one or two (rarely more) sessions. With the exception of 4 patients who underwent embolization with or without surgery before referral, only 1 of our own patients underwent preradiosurgical embolization exclusively. She had a complex bilateral cavernous sinus fistula with such extensive pial venous drainage that adequate angiographic imaging of the fistula itself was impossible beforehand. In other cases, the goal of embolization was to reduce pial or cortical venous drainage, intractable bruit, or ocular/visual symptoms. Polyvinyl alcohol particles were the primary embolization material, and were used in conjunction with silk suture material, gelfoam, and fibered microcoils, as indicated, in the majority of cases. Embolization was successful in reducing (n = 38) or eliminating (n = 2) cortical or pial venous drainage in all cases. However, in all but 1 case, some portion of the fistula still remained angiographically visible after embolization. Thus, the long-term obliteration results described below can be attributed to radiosurgery. Most patients were observed in the hospital overnight after their radiosurgery and embolization sessions. Yearly angiographic follow-up and neurologic assessment was planned. For the most part, we have abandoned earlier and/or more frequent MRI follow-up as being costly and noninformative. We considered additional treatment if complete angiographic obliteration of DAVFs did not occur by 3 years after the initial radiosurgical procedure.

Results

Long-Term Clinical Follow-Up and Management

We reviewed the results of clinical follow-up evaluations for 53 patients between 10 and 84 months after radiosurgical treatment (table 6). No deaths, no fistula-related hemorrhages, and no permanent radiosurgery-induced neurologic deficits have occurred to date in long-term follow-up. Over 96% of patients experienced no posttreatment complications related to radiosurgery, embolization, or other aspects of their disease management. Symptomatic long-term improvement included diminished or absent bruit, relief of intract-

Table 6. Long-term clinical follow-up 10–84 months after radiosurgery and embolization for DAVFs (53 patients treated as of Dec. 31, 1996)

Clinical condition	Patients	
	n	%
Improved	29	55
Stable	14	26
Temporary deficit	2	4
Permanent deficit	1[1]	2
Deficit unrelated to radiosurgery	3[2]	6
Permanent radiosurgery-induced deficit	0	
Hemorrhage or death from DAVF	0	
Death, other causes	1	2
Incomplete follow-up	3	6

[1] Thromboembolic middle cerebral aftery occlusion (see text).
[2] Unrelated hearing loss (n = 1), demyelinating disease (n = 1), contralateral middle cerebral aftery ischemic stroke (n = 1).

able vascular headaches, disappearance of chemosis, proptosis and/or periorbital edema, and/or resolution of extraocular motion deficits.

Because of the dynamic nature of this disorder, some patients experienced fluctuating symptoms or angiographic findings before their fistulas finally obliterated. Two patients required repeat embolization because of symptomatic, fluctuating venous drainage patterns between 1 day and 6 months, respectively, after their initial embolization sessions. Two other patients experienced transient, mild cranial nerve or neurologic deficits after radiosurgery or embolization, respectively. These resolved completely. One other patient developed a left middle cerebral artery thromboembolic occlusion despite systemic heparinization during attempted repeat embolization of a sigmoid sinus DAVF 1 year after her initial treatment. Despite immediate intraarterial thrombolysis, she sustained a permanent neurologic deficit.

Three patients developed new or increased neurologic deficits during follow-up that appeared to be unrelated to treatment of their DAVFs. These included progression of a previously diagnosed demyelinating disorder, a long delayed (10 months) middle cerebral artery distribution ischemic stroke contralateral to the treated transverse sinus DAVF, and transient vertigo with permanent unilateral hearing loss 1 year after radiosurgery and embolization of a transverse sinus DAVF [44]. Careful review of the latter case revealed that the pertinent neurootologic structures were all well outside the radiosurgical treatment field.

Table 7. Long-term angiographic follow-up after radiosurgery and embolization of DAVFs (n = 34)

	Location		
	cavernous sinus	all other locations	total series
Complete obliteration	11	10	21
Near-total obliteration (clinical cure)	1	4	5
Partial obliteration	0	7	7
Unchanged	0	1	1
Proportion of patients cured, n	12/12	14/22	26/34
%	100	64	76

Long-Term Angiographic Obliteration

Follow-up angiography in 34 patients followed for at least 1 year after treatment showed that 26 fistulas (76%) were totally or nearly totally obliterated regardless of location, size, or other factors (table 7). Cases designated as 'near total' obliteration had angiographic findings of only a minor stagnant vein that required no further treatment. These patients were discharged from further scheduled follow-up imaging and are considered clinically cured. Two asymptomatic patients required repeat radiosurgery between 2 and 3 years after their initial treatment. In each case, retrospective review of the original radiosurgical dose plans revealed that a portion of the DAVF had not been covered by the prescribed isodose line. Repeat embolization was performed in 1 case, but was not feasible in the other due to a lack of remaining accessible vessels.

The most interesting finding to emerge from our long-term follow-up is the 100% cure rate for cavernous sinus DAVFs. None of these lesions were judged to be safely curable by direct surgical intervention or endovascular techniques. In contrast, 64% of 1-, 2- or 3-year follow-up angiograms in patients with DAVFs at all other locations combined showed complete obliteration. Whether this finding was related to the larger size, longer segment of involved sinus, more complex geometry, multiplicity and variety of feeding vessels, or flow characteristics of noncavernous sinus DAVFs, remains to be determined.

Illustrative Cases

Case 1 (fig. 1): This 65-year-old man came to the neurosurgical cerebrovascular service at the Mayo Clinic for excision of a 'right temporal lobe arteriovenous malformation'. He was transferred from a rehabilitation unit after having partially recovered from a left hemiparesis as a consequence of hemorrhage from the malformation. Aside from his neurologic findings, physical examination upon hospital admission disclosed a 10-cm abdominal aortic aneurysm (confirmed by imaging studies). Urgent repair was undertaken by the vascular surgery service at which time a synthetic aorto-bifemoral graft was placed.

Thereafter, the patient's outside cerebral angiogram films were analyzed closely in preparation for treatment. The fistulous nature of the malformation was recognized, and given the patient's fragile status, alternatives to conventional open surgery were weighed carefully. Embolization was not a viable option because the major blood supply to the fistula was via the internal carotid artery. Also, the fresh vascular graft precluded femoral angiography, especially for prolonged endovascular procedures. Thus, largely by a process of elimination, this patient became the first case in our dural AV fistula series. Retrograde brachial angiography during imaging for targeting avoided catheterization through the fresh vascular graft. Radiosurgical dosimetry: 18 Gy to the margin at the 50% isodose line using one 8-mm and one 14-mm collimator. As illustrated in figure 1, progressive shrinkage of the draining varix in this high-risk patient occurred over 18 months, and complete obliteration was documented 3 years after radiosurgical treatment. His left hemiparesis improved considerably over time.

Case 2 (fig. 2): This 54-year-old man presented to the neurosurgical service at the Mayo Clinic in Jacksonville, Florida, with the subacute onset of left-sided chemosis, proptosis and extraocular muscle palsies. His visual acuity was not affected. Cerebral angiography showed a left-sided cavernous sinus dural AV fistula, fed predominantly by the internal carotid artery, and drained by an enlarged superior ophthalmic vein. Since embolization of the minor external carotid artery contribution was unlikely to be effective, and since his vision per se was not threatened, the patient was referred for radiosurgical treatment. Dosimetry: 18 Gy to the margin at the 50% isodose line using three 8-mm and two 14-mm collimators. Within 7 months after radiosurgery the patient's ocular symptoms had resolved. Follow-up angiography documented complete obliteration of the fistula 16 months after treatment (fig. 2).

Discussion

Natural History and Rationale for Radiosurgery followed by Embolization

Dural AVFs that present with intractable bruit, intracranial hypertension, progressive visual symptoms, focal neurological deficits, seizures, or hemorrhage warrant treatment. Those with angiographic characteristics associated with a high risk of hemorrhage should also be treated regardless of symptoms [2, 9, 41, 46, 55]. Various studies strongly suggest that the risk of hemorrhage from a

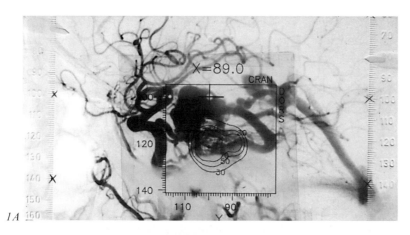

1A

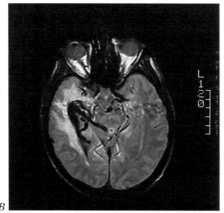

1B

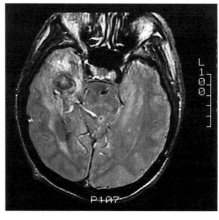

1C

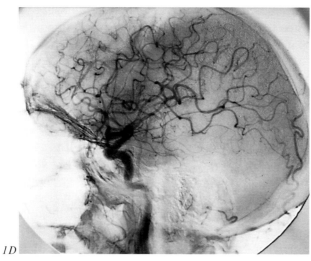

1D

DAVF is related to the location of the nidus and consequently, to the anatomic features of the venous drainage. One review suggested that DAVFs close to dural venous sinuses had a lifetime risk of hemorrhage of 7.5%, whereas those located farther away from a major sinus had a 51% risk of hemorrhage [55]. Other reviews found that retrograde cortical venous drainage, variceal or aneurysmal venous dilations, and/or galenic venous drainage each correlated significantly with an aggressive neurological course [2]. However, as plausible as the foregoing explanations may sound, the frequency with which the features associated with aggressive lesions might also be present in more benign cases remains unknown [49]. The ascertainment bias that leads to the discovery of more apparently symptomatic or 'aggressive' lesions and the selection bias towards reporting the conventional surgical or endovascular treatment of such cases is apparent from a review of the recent literature. Other treatment preference biases may account for the omission of radiosurgery as a management option in recent reviews published well after our series and other smaller case reports had become available in the neurosurgical literature [51].

For selected patients with dural AVFs who are not surgical candidates, we have developed a treatment strategy that includes gamma knife radiosurgery followed in many cases by particulate embolization. The rationale for this approach is that radiosurgery should cause obliteration of most DAVFs within 1–3 years. Embolization after radiosurgery keeps the DAVF angiographically visible for radiosurgical targeting, palliates the patients symptoms, and protects the patient to some extent from early hemorrhage. Even if the embolized portion of the DAVF were to recanalize, the entire lesion is simultaneously undergoing occlusion. We recommend this mode of therapy for DAVFs having direct antegrade venous sinus drainage as well as for those with retrograde venous sinus and pial or cortical drainage, as long as symptoms are localized or mild. Patients who experience a more aggressive clinical course as a consequence of cortical venous hypertension should be treated using embolization and/or surgery. If curative embolization and conventional surgery is not an acceptable option, as was the case for several of our patients, radiosurgery and embolization can provide a useful alternative.

Fig. 1. Case 1: Right retrograde branchiocephalic angiogram (*A*), lateral view, during radiosurgical treatment of a (predominantly) meningohypophyseal trunk-to-basal vein of Rosenthal DAVF that had ruptured a few months before referral. See text for clinical details and dosimetry. Because the predominant blood supply to the fistula was from the internal carotid artery, embolization was *not* performed. Pretreatment (*B*) and 18-month posttreatment (*C*) axial MRI scans illustrate progressive shrinkage and thrombosis of the complex variceal venous drainage. Right carotid angiography 3 years after radiosurgery (*D*) showed complete obliteration of the fistula. Substantial neurologic recovery from deficits caused by the patient's hemorrhage had occurred.

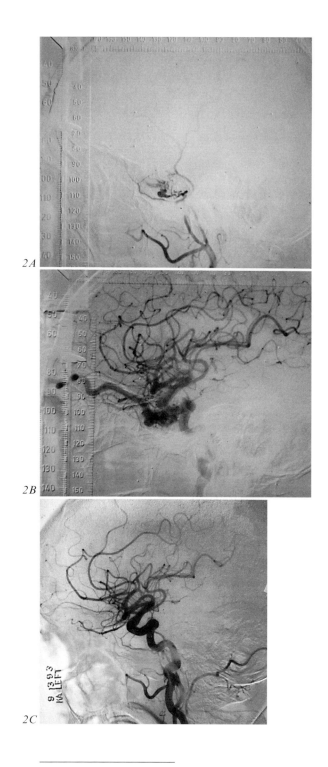

2A

2B

2C

The few early reports of conventional radiation therapy for DAVFs did not include outcome data. Treatment with stereotactic irradiation, usually combined with embolization, also had been reported before our series, but with uncertain results in most cases. Our initial 1996 report described the early results in our first 29 patients. This longer term review reflects the maturation of our patient selection process, the refinement of our management strategy, and the advancement of radiosurgical dose-planning methods. In terms of results, the extraordinary success rate of our treatment strategy for cavernous sinus DAVFs is especially important in light of the disabling visual symptoms (including the threat of blindness), distressing and obvious cosmetic deformity, low likelihood of hemorrhage, and lack of other equally reliable and noninvasive therapy for these lesions. In contrast to fistulas that involve the sagittal, transverse, or sigmoid sinuses, those that involve the cavernous sinus are less amenable to surgical intervention with an acceptable rate of morbidity. While transarterial embolization of the external carotid artery contribution to cavernous sinus DAVFs can provide safe palliation of symptoms, we consider an endovascular approach to the internal carotid artery component as neither feasible nor safe. Transvenous or transorbital venous endovascular approaches to the cavernous sinus remain the treatment of choice at many centers, especially in cases with threatened visual loss due to orbital venous congestion that cannot be managed by external carotid arterial embolization and radiosurgery alone. However, radiosurgery avoids the potential long-term consequences of permanent sinus occlusion.

Review of Clinical Results and Follow-Up Recommendations

The management-related morbidity rate in our series has continued to be low (4% temporary, 2% permanent) and predominantly reflected the need to reduce cortical or pial venous hypertension by using embolization in high-risk patients who were not surgical candidates. Notably, no DAVFs have bled during follow-up after our treatment regimen, including lesions that originally presented with hemorrhage and/or high-risk patterns of venous drainage. In retrospect, the most severe management complication in this series could have been avoided by not intervening further, but rather, by waiting more patiently for slow radiosurgery-induced obliteration to occur instead of performing

Fig. 2. Case 2: Selective left external carotid (*A*) and internal carotid (*B*) angiograms, lateral views, during radiosurgical treatment of a cavernous sinus DAVF. Because the external carotid artery contribution to the fistula was minimal, embolization was *not* performed. See text for clinical details and dosimetry. Left common carotid angiogram (*C*) 16 months after radiosurgery showed complete obliteration of the fistula. The patient's clinic symptoms had resolved entirely by 7 months after treatment.

repeat embolization only 1 year after the original treatment. Other adverse events were either temporary, much less serious, and/or not related to treatment in any way whatsoever (table 6).

We have recommended annual follow-up angiography in most patients who presented with significant symptoms or worrisome angiographic findings. In less symptomatic patients with lesions having a more benign venous drainage pattern, 2-year (and if required, 3-year) follow-up angiography would be sufficient. MRI and MR angiography have turned out to be useful during follow-up in only a few patients where a large draining varix was present and progressive thrombosis or shrinkage could be observed [11]. Angiography has remained necessary to document cure or persistence of the fistula definitively in all cases.

Now that more than 5 years have elapsed since we documented DAVF obliteration in our first patient, no patient has shown any signs of having developed a recurrent or recanalized lesion after documented angiographic cure. Radiosurgery followed by embolization in selected cases is a safe and effective treatment in properly selected patients with symptomatic or high-risk DAVFs. This treatment strategy provides durable results, and may turn out to be the treatment of choice for individuals with DAVFs that involve the cavernous sinus.

References

1   Aminoff MJ, Kendall BE: Asymptomatic dural vascular anomalies. Br J Radiol 1973;46:662–667.
2   Awad IA, Little JR, Akrawi WP, et al: Intracranial dural arteriovenous malformations: Factors predisposing to an aggressive neurological course. J Neurosurg 1990;72:839–850.
3   Barcia-Salorio JL, Soler F, Hernandez G, Barcia JA, et al: Radiosurgical treatment of low flow carotid-cavernous fistulae. Acta Neurochir 1991;52:93–95.
4   Barnwell SL, Halbach VV, Dowd CF, Higashida RT, et al: Dural arteriovenous fistulas involving the inferior petrosal sinus: Angiographic findings in six patients. Am J Neuroradiol 1990;11:511–516.
5   Barnwell SL, Halbach VV, Higashida RT, et al: Complex dural arteriovenous fistulas. Results of combined endovascular and neurosurgical treatment in 16 patients. J Neurosurg 1989;71:352–358.
6   Bitoh S, Sasaki S: Spontaneous cure of dural arteriovenous malformations in the posterior fossa. Surg Neurol 1979;12:111–114.
7   Black P, Uematsu S, Perovic M, et al: Carotid-cavernous fistula: A controlled embolus technique for occlusion of fistula with preservation of carotid blood flow. Technical note. J Neurosurg 1973;38:113–118.
8   Borden JA, Wu JK, Shucart WA: A proposed classification for spinal and cranial dural arteriovenous fistulous malformations and implications for treatment. J Neurosurg 1995;82:166–179.
9   Brown RD, Wiebers DO, Nichols DA: Intracranial dural arteriovenous fistulae: Angiographic predictors of intracranial hemorrhage and clinical outcome in nonsurgical patients. J Neurosurg 1994;81:531–538.
10 Chandler HC Jr, Friedman WA: Successful radiosurgical treatment of a dural arteriovenous malformation: Case report. Neurosurgery 1993;33:139–142.
11 Chen J-C, Tsuruda JS, Halbach VV: Suspected dural arteriovenous fistula: Results of screening MR angiography in seven patients. Radiology 1992;183:265–271.
12 Costin JA, Weinstein MA, Berlin AJ, et al: Dural arterio-venous malformations involving the cavernous sinus: A case report. Br J Ophthalmol 1978;62:478–482.

13   Courtheoux P, Labbe D, Hamel C, et al: Treatment of bilateral spontaneous dural carotid-cavernous fistulas by coils and sclerotherapy. Case report. J Neurosurg 1987;66:468–470.

14   Davies MA, TerBrugge K, Willinsky R, et al: The validity of classification for the clinical presentation of intracranial dural arteriovenous fistulas. J Neurosurg 1996;85:830–837.

15   Eisenman JI, Alekoumbides A, Pribram H: Spontaneous thrombosis of vascular malformations of the brain. Acta Radiol (Diagn) 1972;13:77–85.

16   Fabrikant JI, Lyman JT, Hosobuchi Y: Stereotactic heavy-ion Bragg peak radiosurgery for intracranial vascular disorders. Methods for treatment of deep arteriovenous malformations. Br J Radiol 1984;57:479–490.

17   Fardoun R, Adam Y, Mercier P, et al: Tentorial arteriovenous malformation presenting as an intracerebral hematoma. Case Report. J Neurosurg 1981;55:976–978.

18   Fermand M, Reizine D, Melki JP, et al: Long-term follow-up of 43 pure dural arteriovenous fistulae of the lateral sinus. Neuroradiology 1987;29:348–353.

19   Gobin YP, Rogopoulos A, Aymard A, et al: Endovascular treatment of intracranial dural arteriovenous fistulas with spinal perimedullary venous drainage. J Neurosurg 1992;77:718–723.

20   Grisoli F, Vincentelli F, Fuchs S, et al: Surgical treatment of tentorial arteriovenous malformations draining into the subarachnoid space. Report of four cases. J Neurosurg 1984;60:1059–1066.

21   Grossman RI, Sergott RC, Goldberg HI, et al: Dural malformations with ophthalmic manifestations: Results of particulate embolization in seven patients. Am J Neuroradiol 1985;6:809–813.

22   Halbach VV, Higashida RT, Hieshima GB, et al: Dural fistulas involving the transverse and sigmoid sinuses: Results of treatment in 28 patients. Radiology 1987;163:443–447.

23   Halbach VV, Higashida RT, Hieshima GB, et al: Dural fistulas involving the cavernous sinus: Results of treatment in 30 patients. Radiology 1987;163:437–442.

24   Halbach VV, Higashida RT, Hieshima GB, et al: Treatment of dural fistulas involving the deep cerebral venous system. Am J Neuroradiol 1989;10:393–399.

25   Hamada Y, Goto K, Inoue T, et al: Histopathological aspects of dural arteriovenous fistulas in the transverse-sigmoid sinus region in nine patients. Neurosurgery 1997;40:452–458.

26   Hamby WB: Carotid-cavernous fistula: Report of 32 surgically treated cases and suggestions for definitive operation. J Neurosurg 1964;21:859–865.

27   Hansen JH, Søgaard IB: Spontaneous regression of an extra- and intracranial arteriovenous malformation. J Neurosurg 1976;45:338–341.

28   Hardy RW, Costin JA, Weinstein M, et al: External carotid cavernous fistula treated by transfemoral embolization. Surg Neurol 1978;9:255–256.

29   Herman JM, Spetzler RF, Bederson JB, et al: Genesis of a dural arteriovenous malformation in a rat model. J Neurosurg 1995;83:539–545.

30   Hosobuchi Y, Fabrikant J, Lyman JT: Stereotactic heavy-particle irradiation of intracranial arteriovenous malformations. Appl Neurophysiol 1987;50:248–252.

31   Houser OW, Baker HL Jr, Rhoton AL Jr, et al: Intracranial dural arteriovenous malformations. Radiology 1972;105:55–64.

32   Houser OW, Campbell JK, Campbell RJ, et al: Arteriovenous malformation affecting the transverse dural venous sinus – An acquired lesion. Mayo Clin Proc 1979;54:651–661.

33   Hugosson R, Bergström K: Surgical treatment of dural arteriovenous malformation in the region of the sigmoid sinus. J Neurol Neurosurg Psychiatry 1974;37:97–101.

34   Ishikawa T, Houkin K, Tokuda K, et al: Development of anterior cranial fossa dural arteriovenous malformation following head trauma. Case report. J Neurosurg 1997;86:291–293.

35   Ishimori S, Hattori M, Shibata Y, et al: Treatment of carotid-cavernous fistula by gelfoam embolization. J Neurosurg 1967;27:315–319.

36   Ito M, Sonokawa T, Mishina H, Sato K: Reversible dural arteriovenous malformation-induced venous ischemia as a cause of dementia: Treatment by surgical occlusion of draining dural sinus. Case report. Neurosurgery 1995;37:1187–1192.

37   Iwama T, Hashimoto N, Takagi Y, et al: Hemodynamic and metabolic disturbances in patients with intracranial dural arteriovenus fistulas: Position emission tomography evaluation before and after treatment. J Neurosurg 1997;86:806–811.

38 Kataoka K, Taneda M: Angiographic disappearance of multiple dural arteriovenous malformations. Case report. J Neurosurg 1984;60:1275–1278.

39 Kendall B: Percutaneous embolic occlusion of the dural arterio-venous malformation. Br J Radiol 1973;46:520–523.

40 Kerber CW, Newton TH: The macro- and microvasculature of the dura mater. Neuroradiology 1973;6:175–179.

41 King WA, Martin NA: Intracerebral hemorrhage due to dural arteriovenous malformations and fistulae. Neurosurg Clin North Am 1992;3:577–590.

42 Kobayashi H, Hayashi M, Noguchi Y, et al: Dural arteriovenous malformations in the anterior cranial fossa. Surg Neurol 1988;30:396–401.

43 Kupersmith MJ, Berenstein A, Choi IS, et al: Management of nontraumatic vascular shunts involving the cavernous sinus. Ophthalmology 1988;95:121–130.

44 Landman JA, Braun IF: Spontaneous closure of a dural arteriovenous fistula associated with acute hearing loss. AJNR 1985;6:448–449.

45 Lasjaunias P, Berenstein A: Surgical Neuroangiography. 2. Endovascular Treatment of Craniofacial Lesions. Berlin, Springer, 1987.

46 Lasjaunias P, Chui M, TerBrugge K: Neurological manifestations of intracranial dural arteriovenous malformations. J Neurosurg 1986;64:724–730.

47 Lawton MT, Jacobowitz R, Spetzler RF: Redefined role of angiogenesis in the pathogenesis of dural arteriovenous malformations. J Neurosurg 1997;87:267–274.

48 Lewis AI, Tomsick TA, Tew JM Jr: Management of tentorial arteriovenous malformations: Transarterial embolization combined with stereotactic radiation or surgery. J Neurosurg 1994;81:851–859.

49 Link MJ, Coffey RJ, Nichols DA, et al: The role of radiosurgery and particulate embolization in the treatment of dural arteriovenous fistulas. J Neurosurg 1996;84:804–809.

50 Lucas CP, Oliveira ED, Tedeschi H, et al: Sinus skeletonization: A treatment for dural arteriovenous malformations of the tentorial apex. J Neurosurg 1996;84:514–517.

51 Lucas CP, Zabramski JM, Spetzler RF, Jacobowitz R: Treatment for intracranial dural arteriovenous malformations: A meta-analysis from the English language literature. Neurosurgery 1997;40:1119–1132.

52 Lunsford LD, Kondziolka D, Flickinger JC, Bisonette DJ, Jungreis CA, Maitz AH, Horton JA, Coffey RJ: Stereotactic radiosurgery for arteriovenous malformations of the brain. Neurosurgery 1991;75:512–524.

53 Magidson MA, Weinberg PE: Spontaneous closure of a dural arteriovenous malformation. Surg Neurol 1976;6:107–110.

54 Mahalley MS Jr, Boone SC: External carotid-cavernous fistula treated by arterial embolization. Case report. J Neurosurg 1974;40:110–114.

55 Malik GM, Pearce JE, Ausman JI, et al: Dural arteriovenous malformations and intracranial hemorrhage. Neurosurgery 1984;15:332–339.

56 Manaka S, Izawa M, Nawata H: Dural arteriovenous malformation treated by artificial embolization with liquid silicone. Surg Neurol 1977;7:63–65.

57 Manelfe C, Berenstein A: Traitement des fistules carotido-caverneuses par voie veineuse. A propos d'un cas. J Neuroradiol 1980;7:13–19.

58 Mendelowitsch A, Gratzl O, Radü EW: Current therapeutic methods of dural arteriovenous malformation: Are there any alternatives? Two case reports of infratentorial AVMs. Neurosurg Rev 1989;12:141–145.

59 Morita A, Meyer FB, Nichols DA, Patterson MC: Childhood dural arteriovenous fistulae of the posterior dural sinuses: Three case reports and literature review. Neurosurgery 1995;37:1193–1200.

60 Mullan S, Johnson DL: Combined sagittal and lateral sinus dural fistulae occlusion. J Neurosurg 1995;82:159–165.

61 Mullan S: Reflections upon the nature and management of intracranial and intraspinal vascular malformations and fistulae. J Neurosurg 1994;80:606–616.

62 Mullan S: Treatment of carotid-cavernous fistulas by cavernous sinus occlusion. J Neurosurg 1979;50:131–144.

63   Newton TH, Cronqvist S: Involvement of the dural arteries in intracranial arteriovenous malforma-
      tions. Radiology 1969;93:1071–1078.
64   Olutola PS, Eliam M, Molot M, et al: Spontaneous regression of a dural arteriovenous malformation.
      Nerosurgery 1983;12:687–690.
65   Pang D, Kerber C, Biglan AW, Ahn HS: External carotid-cavernous fistula in infancy: Case report
      and review of the literature. Neurosurgery 1981;8:212–218.
66   Peeters FLM, Kröger R: Dural and direct cavernous sinus fistulas. AJR 1979;132:599–606.
67   Pierot L, Chiras J, Meder JF, Rose M, Rivierez M, Marsault C: Dural arteriovenous fistulas of the
      posterior fossa draining into subarachnoid veins. Am J Neuroradiol 1992;13:315–323.
68   Pugatch RD, Wolpert SM: Transfemoral embolization of an external carotid-cavernous fistula. Case
      report. J Neurosurg 1975;42:94–97.
69   Roy D, Raymond J: The role of transvenous embolization in the treatment of intracranial dural
      arteriovenous fistulas. Neurosurgery 1997;40:1133–1144.
70   Sano H, Jain VK, Kato Y, Tanji H, Kanno T, Adachi K, Katada K: The treatment of dural AVM
      by embolization with aron alpha-(ethyl-2-cyanoacrylate). Acta Neurochir (Wien) 1987;88:10–19.
71   Steiner L, Lindquist C, Adler JR, Torner JC, Alves W, Steiner M: Clinical outcome of radiosurgery
      for cerebral arteriovenous malformations. J Neurosurg 1992;77:1–8.
72   Sundt TM Jr, Piepgras DG: The surgical approach to arteriovenous malformations of the lateral
      and sigmoid dural sinuses. J Neurosurg 1983;59:32–39.
73   Takahashi A, Yoshimoto T, Kawakami K, Sugawara T, Suzuki J: Transvenous copper wire insertion
      for dural arteriovenous malformations of cavernous sinus. J Neurosurg 1989;70:751–754.
74   Teng MM, Guo WY, Huang CI, Wu CC, Chang T: Occlusion of arteriovenous malformations of
      the cavernous sinus via the superior ophthalmic vein. Am J Neuroradiol 1988;9:539–546.
75   Terada T, Higashida RT, Halbach VV, et al: Development of acquired arteriovenous fistulas in rats
      due to venous hypertension. J Neurosurg 1994;80:884–889.
76   Thompson BG, Doppman JL, Oldfield EH: Treatment of cranial dural arteriovenous fistulae by
      interruption of leptomeningeal venous drainage. J Neurosurg 1994;80:617–623.
77   Tsai FY, Hieshima GB, Mehringer CM, Grinnel V, Pribram HW: Delayed effects in the treatment
      of carotid-cavernous fistulas. Am J Neuroradial 1983;4:357–361.
78   Tu YK, Liu HM, Hu SC: Direct surgery of carotid cavernous fistulae and dural arteriovenous
      malformations of the cavernous sinus. Neurosurgery 1997;41:798–806.
79   Uflacker R, Lima S, Ribas GC, Piske RL: Carotid-cavernous fistulas: Embolization through the
      superior ophthalmic vein approach. Radiology 1986;159:175–179.
80   Vidyasagar C: Persistent embryonic veins in the arteriovenous malformations of the dura. Acta
      Neurochir 1979;48:199–216.
81   Vinuela F, Fox AJ, Debrun GM, Peerless SJ, Drake CG: Spontaneous carotid-cavernous fistulas:
      Clinical, radiological, and therapeutic considerations. Experience with 20 cases. J Neurosurg 1984;
      60:976–984.
82   Vinuela F, Fox AJ, Pelz DM, Drake CG: Unusual clinical manifestations of dural arteriovenous
      malformations. J Neurosurg 1986;64:554–558.
83   Vinuela FV, Debrun GM, Fox AJ, Kan S: Detachable calibrated-leak balloon for superselective
      angiography and embolization of dural arteriovenous malformations. J Neurosurg 1983;58:817–823.
84   Voigt K, Sauer M, Dichgans J: Spontaneous occlusion of a bilateral caroticocavernous fistula studied
      by serial angiography. Neuroradiology 1971;2:207–211.
85   Yoshimura S, Hashimoto N, Kazekawa K, et al: Arteriovenous fistula around the ventriculoperi-
      toneal shunt system in a patient with a dural arteriovenous malformation of the posterior fossa.
      Case report. J Neurosurg 1995;82:288–290.

Bruce E. Pollock, MD, Department of Neurologic Surgery, Mayo Clinic,
200 First Street, SW, Rochester, MN 55905 (USA)
Tel. (507) 284 2775, Fax (507) 284 5206

Lunsford LD, Kondziolka D, Flickinger JC (eds): Gamma Knife Brain Surgery.
Prog Neurol Surg. Basel, Karger, 1998, vol 14, pp 78–88

....

Stereotactic Radiosurgery for Cavernous Malformations

Douglas Kondziolka, L. Dade Lunsford, John C. Flickinger

Departments of Neurological Surgery and Radiation Oncology and
The Center for Image-Guided Neurosurgery, University of Pittsburgh, Pa., USA

Few concepts in radiosurgery stimulate as much discussion as the management of cerebral cavernous malformations. The lack of an imaging test to confirm obliteration, and the lack of a flow-based imaging method to assist vessel targeting present unresolved challenges. In 1990, we reported our radiosurgery experience in 24 patients [10]. Since that time both patient selection criteria and the techniques of treatment planning and dose selection have evolved. By the early 1990s, even microsurgical resection was reported only for small series of patients. Most physicians recommended continued observation for patients who sustained a brain hemorrhage from a cavernous malformation in a critical location. We hypothesized that the vessels of angiographically occult vascular malformations (most are cavernous malformations) would respond to ionizing radiation in a similar way that the blood vessels of an arteriovenous malformation [14]. We anticipated that radiosurgery would reduce the hemorrhage risk in individual patients after a latency interval [2].

The clinical benefit of cavernous malformation radiosurgery must be confirmed by a reduced hemorrhage rate, and must be accompanied by low treatment morbidity. In carefully selected patients who had sustained at least two prior symptomatic bleeds, we identified that the annual hemorrhage rate dropped significantly approximtely two years after radiosurgery [13]. Critics of this technique have argued that cavernous malformations may have periods of increased or decreased hemorrhagic activity, regardless of whether or not radiosurgery was performed. Our understanding of the natural history of these lesions contributes to our selection of individual management strategies, and has assisted in the design of a prospective randomized trial [4, 12, 15, 18, 25].

The Imaging Appearance of Cavernous Malformations

The use of magnetic resonance imaging (MRI) led to a significant increase in the diagnosis of intracerebral cavernous malformations. MRI identified those malformations in lobar locations and those deep lesions that presented on a pial or ependymal surface, thereby making them suitable for resection. It is likely that the majority of angiographically occult vascular malformations are cavernous-type malformations [16, 19]. We performed high-resolution subtraction angiography to determine whether each vascular malformation was truly occult. Patients with venous malformations identified either on MRI or angiography were not accepted for any kind of treatment, whether radiosurgery or resection [9]. Increasingly we have identified patients that have cavernous malformations in association with venous malformations [3, 17]. In the clinical context of observed hemorrhage, it is most likely that the cavernous malformation has bled [17]. We hypothesized that regional venous hypertension caused by the venous malformation may lead to repeated microhemorrhage and cavernous malformation expansion [3]. We performed radiosurgery in 1 patient who had a thalamic cavernous malformation (and adjacent venous malformation) with two hemorrhages. In some patients it can be difficult to sort out the nature of the malformation(s) based on imaging alone.

Microsurgery or Radiosurgery?

During the 10-year interval of this radiosurgery study, several groups reported the successful and relatively safe removal of deep-seated cavernous malformations. Kashiwagi et al. [8] reported 4 patients who had resection of their brainstem cavernous malformations; only 1 had additional morbidity. Zimmerman et al. [28] reported 24 patients with brainstem cavernous malformations, 16 of whom had surgery. Four patients had transient new deficits and 12 were the same or improved in the immediate postoperative period. Seven of the 8 patients who were managed conservatively remained stable, although 1 died from rehemorrhage.

Bertalanffy et al. [1] reported 26 patients who had microsurgery for deep cavernous malformations. Eighteen patients had surgery because of progressive neurologic deficits; 15 had had a hemorrhage. Successful total resection without morbidity was achieved in 11 patients but 7 had new postoperative deficits. An additional 8 patients had residual major deficits, 5 had lesions in the basal ganglia, thalamus, or insula. Based on their unsatisfactory results, these authors questioned whether the indications for deep cavernous malformation surgery should be limited [1]. Other groups however, emphasized the benefits of intraoperative localization followed by meticulous microsurgical technique to attain a good surgical outcome in deep brain locations [5–7, 20–24]. These reports provided support that at experienced centers, some cavernous malformations can be totally resected with low risk for permanent morbidity, provided that the malformation or hematoma presented to a pial or ependymal surface (brainstem or thalamus). The problems posed by surgical resection prompted the development of an additional therapeutic strategy.

The Rationale for Radiosurgery

In 1988 we began our use of stereotactic radiosurgery as an alternative management approach. We speculated that the vessels of a cavernous malformation might respond similarly as an arteriovenous malformation (AVM) to high, single-session radiation doses. This response might consist of endothelial cell proliferation, vessel wall hyalinization, and eventual luminal closure. A time period of 2–3 years might be necessary to achieve this result. Radiosurgery was recommended only for intraparenchymal malformations in deep brain locations where the risks of surgical resection were felt to be excessive. At that time, the successful resection of deeply located cavernous malformations with low morbidity was reported in case reports or small series [6, 20, 22, 26, 27]. Without appreciable surgical experience that detailed successful, safe removal of these lesions, most patients even after hemorrhage were managed conservatively. We declined to recommend radiosurgery to patients who had minimal symptoms and who had sustained only one hemorrhage. Rather, we recommended either microsurgical resection (depending upon location) or further observation with periodic imaging studies. The latter patients were entered into a separate prospective natural history study that was reported in 1995. That study found that patients without a prior symptomatic bleed had a 0.6% annual hemorrhage rate, and those with one prior hemorrhage, a rate of 4.5% (irrespective of brain location) [12]. A recent report by Porter et al. [15] found that deep location was significant for the development of new neurologic symptoms from cavernous malformations.

University of Pittsburgh Experience

Over a 10-year interval, we performed stereotactic radiosurgery on 63 patients with cavernous malformations. The mean patient age was 38 years. There were 33 males and 30 females. The median number of hemorrhages before radiosurgery was two per patient. Fifty-nine patients had at least two

Table 1. Location of cavernous malformations for radiosurgery (n = 63)

Location	Patients, n
Pons/midbrain	36
Medulla	3
Thalamus	9
Basal ganglia	5
Temporal lobe	4
Parietal lobe	6

hemorrhages (range 2–9); 4 had cavernous malformations in critical locations and had sustained only one documented hemorrhage followed by an additional stepwise decline in neurological function. Ten patients (16%) had undergone attempts at surgical resection of their malformation. All patients had MRI and most had angiography before radiosurgery. The MRI appearance was typical of a cavernous malformation in each patient [16]. The brain locations of cavernous malformations in this series are shown in table 1. No patient had a seizure disorder.

Our technique for cavernous malformation radiosurgery begins with application of the Leksell model G stereotactic frame (Elekta Instruments, Atlanta, Ga., USA) under local scalp infiltration anesthesia and mild oral or intravenous sedation. Stereotactic MRI is then performed (computed tomography (CT) planning was used before 1990). After a sagittal scout acquisition is performed (short TR), axial short and long TR images are obtained at 3-mm image intervals. An additional contrast-enhanced axial spoiled-grass volume acquisition (512 × 256 matrix) is used to obtain images at 1-mm intervals. Although we acknowledge that definition of the cavernous malformation 'nidus' is problematic, we define the malformation as the region characterized by mixed signal change within an outer hemosiderin ring, typified by low signal intensity. Hematoma eccentric from the malformation is excluded from dose planning. Single or multiple isocenter plans are used to construct a conformal irradiation volume for the cavernous malformation margin (fig. 1). In all patients, the 50% isodose or greater is used for the margin. Dose planning and selection is performed by a neurosurgeon, radiation oncologist and medical physicist. Selection of dose for radiosurgery is dependent upon malformation volume [10]; we reduce the dose below that advocated for an AVM of similar volume. Radiosurgery is performed with a 201-source cobalt-60 Gamma knife (Elekta Instruments). After radiosurgery, all patients receive methylprednisolone, 40 mg intravenously. Patients are discharged from hospital the next morning. Follow-up imaging studies are requested at 6-month intervals for the first

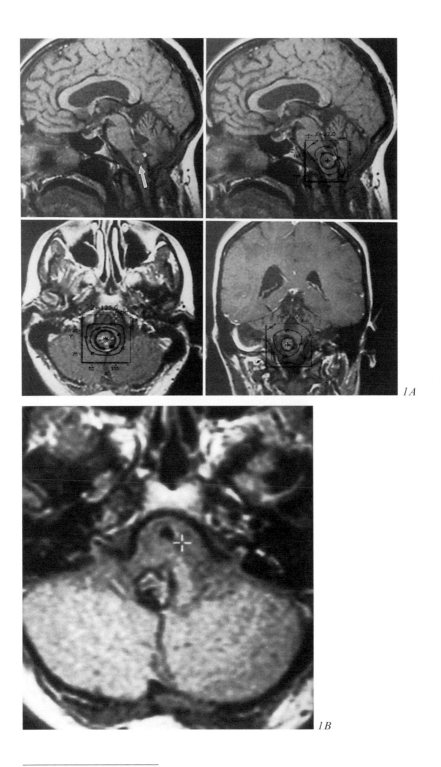

1A

1B

2 years after radiosurgery, yearly thereafter for 2 years, and then biannually (fig. 2).

The mean malformation volume was 1.7 ml (range 0.12–8.1). An average of 2.8 isocenters was used for dose planning (range 1–8). The mean dose delivered to the margin of the malformation was 16 Gy.

Reduction of Hemorrhage Rate after Radiosurgery

We reported a detailed evaluation of the first 47 patients who were followed through the pre- and postradiosurgery observation periods [13]. Before radiosurgery in this group, there were 193 patient-years of observation (calculated from date of first hemorrhage in each patient). Forty-four patients sustained at least two hemorrhages, and 3 patients had one hemorrhage. The mean observation interval before radiosurgery was 4.12 years (range 0.5–12). The occurrence of 109 bleeds in the 193 years of observation led to an annual hemorrhage rate of 56.5%. If the first bleed was used only to identify the malformation and was excluded from analysis, then the subsequent annual symptomatic bleed rate in these patients was 32% [13]. The lifetime risk since birth was 5.9% per year.

The posttreatment observation period was from the time of radiosurgery until last follow-up, surgery, or death (fig. 3). In this series, follow-up after radiosurgery varied from 0.33 to 6.4 years (mean 3.6). The shortest follow-up (0.33 years) was in a patient who sustained a hemorrhage after radiosurgery and underwent resection. After irradiation, a total of eight hemorrhages were observed in 169 patient-years of observation. Because of an estimated 2-year latency interval for maximum vascular effects, we stratified results from 0 to 2 years after treatment, and from 2 to 6 years. From 0 to 2 years after radiosurgery, seven bleeds in 80 observation years were found (annual hemorrhage rate of 8.8%). From 2 to 6 years, only one bleed in 89 observation years was found, for an annual hemorrhage rate of 1.1%. Only 6 of these 47 patients bled after radiosurgery (McNemar $\chi^2 = 36$, p < 0.0001) [13]. No hemorrhages have occurred in this patient group in the additional 2 years of follow-up since our 1995 report.

Fig. 1. A Sagittal and axial MR scans at radiosurgery showing a cavernous malformation of the medulla (arrow). This woman had multiple bleeds that had necessitated ventilation. Prior to MRI, a diagnosis of multiple sclerosis had been suggested. The radiosurgery isodose plan using one isocenter is shown. *B* Two years later, the lesion and brainstem remained stable. She has not had a hemorrhage in the 8 years since radiosurgery.

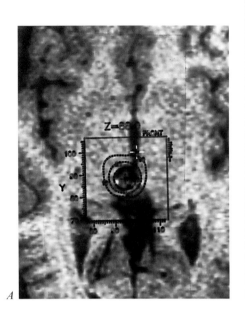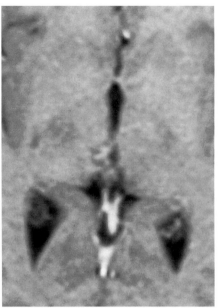

A

B

Fig. 2. A Axial MR scan at radiosurgery showing a cavernous malformation in the right thalamus. One 8-mm isocenter was used. *B* Her MR scan 4 years later remained unchanged.

Adverse Effects of Radiosurgery

For vascular malformations or tumors, adverse radiation effects can be observed within the first 3 years of irradiation. Within 3–18 months after cavernous malformation radiosurgery, 12 of the first 47 patients developed new neurologic deficits that were associated with parenchymal imaging changes (regions of increased signal on long TR MR images surrounding the lesion) (26%). Neurologic symptoms were location-dependent. Two patients with thalamic cavernous malformations developed hemiparesis requiring prolonged oral corticosteroid use before recovery. In 8 patients, deficits were temporary with full recovery. Two patients were left with permanent neurologic deficits from radiosurgery (4%); both patients remained functional with Karnofsky performance scores of 80 and 90 [13]. The risk of radiosurgery for cavernous malformations may be greater than the risks for other pathologic entities, although this remains poorly defined. The critical locations of these malformations may be the explanation for the observed increased morbidity as compared to our general experience with AVMs or tumors. In comparison to our brainstem AVM series (for which we found a temporary morbidity rate of 16%

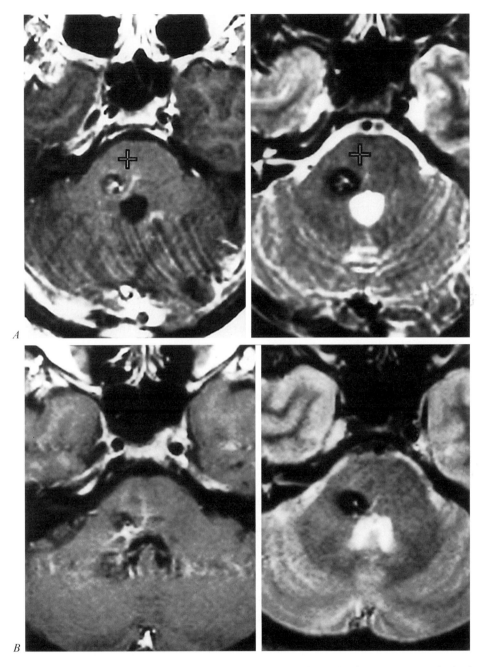

Fig. 3. A Axial MR images of a pontine cavernous malformation in a young man who had sustained two symptomatic hemorrhages. *B* 18 months after radiosurgery, he remained well with a stable appearance of the malformation. An adjacent venous malformation can be seen.

[11]), the 26% temporary morbidity rate in this cavernous malformation series was higher. In the 23 patients managed after our recommendation of a reduced dose, we did not identify a reduced total morbidity rate although the severity of complications was reduced.

Imaging Changes after Radiosurgery

In most patients, serial MRI studies showed no change in the malformation with stable appearance in volume, hemosiderin, or the central mixed signal. Ten of the first 47 patients had a reduction in the size of their malformation, which either represented a treatment-related effect on malformation vessels, or continued resorption of hematoma/microhemorrhage. We do not interpret malformation regression on imaging as a sign of 'obliteration'. Because there is no imaging test that can confirm vessel obliteration, we cannot report 'complete obliteration' of a cavernous malformation just because the patient survived several years after treatment without sustaining a new hemorrhage. We believe that new imaging tools may be able to detect patent or closed vascular spaces within these malformations, perhaps through a quantification of blood volume.

A Prospective Randomized Trial of Radiosurgery versus Conservative Management for High-Risk Cavernous Malformations

Together with members of the Joint Section on Cerebrovascular Disease of the American Association of Neurological Surgeons and the Congress of Neurological Surgeons, we designed a randomized, multicenter prospective trial. Patients will be entered over a 3-year period and the last patients followed for a minimum of 3 years. Apart from the treatment itself, which is a one-time intervention, patients in both arms of the study will be managed identically. Those randomized to stereotactic radiosurgery will undergo the procedure as soon as possible. Diagnostic evaluations will consist of the following: (a) patients will be evaluated at 6, 12, 18, 24 and 36 months posttreatment and then yearly. Results will be used to score the NIH stroke scale. (b) Functional disability and quality-of-life assessments will be done using a quality-of-life questionnaire developed and validated at Yale University. (c) MRI scanning will be performed at the follow-up intervals and as needed for patients with new neurologic problems. Inclusion criteria include: (a) patients who have had two neurologic events in the preceding 3 years (neurologic event equals sudden onset of new neurologic symptom or sign lasting a minimum of 3 months

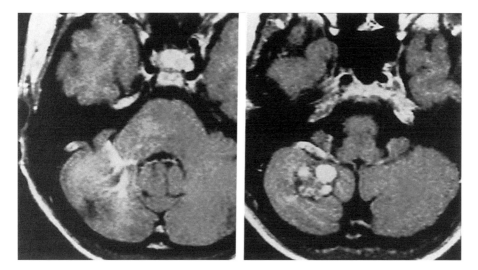

Fig. 4. MR images in a young woman who had one hemorrhage from a cerebellar cavernous malformation. She made a full recovery. An adjacent venous malformation is shown (left). She had been managed conservatively and is followed with serial imaging studies.

directly attributed to a lesion on MRI) (fig. 4); (b) the patient must have a lesion typical of a cavernous malformation on MRI; (c) the cavernous malformation must be < 2.5 cm. in maximal diameter, and (d) the lesion must be located in a region of high surgical risk (this decision will be left up to the discretion of the participating center and adjudicated by an independent committee). In general, patients should be considered for entry into the study if more than half of the lesion is the brainstem, basal ganglia or thalamus. This study has begun and is planned for a 6-year period. We hope to clarify the indications for, and expectations after cavernous malformation radiosurgery.

References

1 Bertalanffy H, Gilsbach IM, Eggert HR, et al: Microsurgery of deep-seated cavernous angiomas: Report of 26 cases. Acta Neurochir (Wien) 1991;108:91–99.
2 Coffey RJ, Lunsford LD: Radiosurgery of cavernous malformations and other angiographically occult vascular malformations; in Awad AI, Barrow DL (eds): Cavernous Malformations. Park Ridge/Ill, American Association of Neurological Surgeons, 1993, pp 187–200.
3 Comey C, Kondziolka D, Yonas H: Regional parenchymal enhancement with mixed cavernous/venous malformations of the brain. J Neurosurg 1997;86:154–158.
4 Del Curling O, Kelly DL, Elster AD, et al: An analysis of the natural history of cavernous angiomas. J Neurosurg 1991;75:702–708.

5   Fahlbusch R, Strauss C, Huk W, et al: Surgical removal of pontomesencephalic cavernous hemangiomas. Neurosurgery 1990;26:449–457.
6   Giombini S, Morello G: Cavernous angiomas of the brain. Acount of fourteen personal cases and review of the literature. Acta Neurochir 1978;40:61–82.
7   Isamat F, Conesa G: Cavernous angiomas of the brain stem. Neurosurg Clin North Am 1993;4:507–518.
8   Kashiwagi S, Van Loveren HR, Tew JM, et al: Diagnosis and treatment of vascular brainstem malformations. J Neurosurg 1990;72:27–34.
9   Kondziolka D, Dempsey PK, Lunsford LD: The case for conservative management of venous angiomas. Can J Neurol Sci 1991;18:295–299.
10 Kondziolka D, Lunsford LD, Coffey RJ, et al: Stereotactic radiosurgery of angiographically occult vascular malformations: Indications and preliminary experience. Neurosurgery 1990;27:892–900.
11 Kondziolka D, Lunsford LD, Flickinger JC: Intraparenchymal brain stem radiosurgery. Neurosurg Clin North Am 1993;4:469–479.
12 Kondziolka D, Lunsford LD, Kestle JR: The natural history of cerebral cavernous malformations. J Neurosurg 1995;83:820–824.
13 Kondziolka D, Lunsford LD, Flickinger JC, et al: Reduction of hemorrhage risk after stereotactic radiosurgery for cavernous malformations. J Neurosurg 1995;83:825–831.
14 Lunsford LD, Kondziolka D, Flickinger JC, et al: Stereotactic radiosurgery for arteriovenous malformations of the brain. J Neurosurg 1991;75:512–524.
15 Porter P, Willinsky R, Harper W, et al: Cerebral cavernous malformations: Natural history and prognosis after clinical deterioration with or without hemorrhage. J Neurosurg 1997;87:190–197.
16 Rigamonti D, Drayer BP, Johnson PC, et al: The MRI appearance of cavernous malformations (angiomas). J Neurosurg 1987;67:518–524.
17 Rigamonti D, Spetzler RF: The association of venous and cavernous malformations. Report of four cases and discussion of the pathophysiological, diagnostic, and therapeutic implications. Acta Neurochir (Wien) 1988;92:100–105.
18 Robinson JR, Awad IA, Little JR: Natural history of the cavernous angioma. J Neurosurg 1991;75:709–714.
19 Robinson JR, Awad IA, Masaryk TJ, et al: Pathological heterogeneity of angiographically occult vascular malformations of the brain. Neurosurgery 1993;33:547–555.
20 Roda JM, Alverez F, Isla A, et al: Thalamic cavernous malformations. Case report. J Neurosurg 1990;72:647–649.
21 Scott RM, Barnes P, Kupsky W, et al: Cavernous angiomas of the central nervous system in children. J Neurosurg 1992;76:38–46.
22 Scott BB, Seeger JF, Schneider RC: Successful evacuation of pontine hematoma secondary to rupture of a pathologically diagnosed 'cryptic' vascular malformations. J Neurosurg 1973;39:104–108.
23 Seifert V, Gaab MR: Laser-assisted microsurgical extirpation of a brainstem cavernoma: Case report. Neurosurgery 1989;25:986–990.
24 Shah MV, Heros RC: Microsurgical treatment of supratentorial lesions; in Awad AI, Barrow DL (eds): Cavernous Malformations. Park Ridge/Ill, American Association of Neurological Surgeons, 1993, pp 101–116.
25 Tung H, Giannotta SL, Chandrasoma PT, et al: Recurrent intraparenchymal hemorrhages from angiographically occult vascular malformations. J Neurosurg 1990;73:174–180.
26 Vaquero J, Salazar J, Martinez T: Cavernomas of the central nervous system: Clinical syndromes, CT scan diagnosis, and prognosis after surgical treatment. Acta Neurochir (Wien) 1987;85:29–33.
27 Yoshimoto T, Suzuki J: Radical surgery on cavernous angioma of the brainstem. Surg Neurol 1986;26:72–78.
28 Zimmerman RS, Spetzler RF, Lee KS, et al: Cavernous malformations of the brain stem. J Neurosurg 1991;75:32–39.

Douglas Kondziolka, MD, University of Pittsburgh Medical Center, Suite B-400,
Department of Neurological Surgery, 200 Lothrop Street, Pittsburgh, PA 15213 (USA)
Tel. (412) 647 6782, Fax (412) 647 0989

Lunsford LD, Kondziolka D, Flickinger JC (eds): Gamma Knife Brain Surgery.
Prog Neurol Surg. Basel, Karger, 1998, vol 14, pp 89–103

····················

Gamma Knife Radiosurgery for Acoustic Neuromas

L. Dade Lunsford [a], *Douglas Kondziolka* [a], *Bruce E. Pollock* [b],
John C. Flickinger [a]

[a] University of Pittsburgh Medical Center, Pittsburgh, Pa., and
[b] Department of Neurologic Surgery, Mayo Clinic, Rochester, Minn., USA

For patients with newly diagnosed or recurrent acoustic nerve sheath tumors (vestibular schwannomas), selection of management options often presents a therapeutic dilemma. Fortunately, simultaneous improvements in surgical and radiosurgical strategies have greatly improved the outlook for patients with acoustic neuromas. The widespread availability of high resolution magnetic resonance imaging (MRI) solved the diagnostic queries that previously obscured therapeutic decision-making. Patients even with minimal symptoms such as imbalance, tinnitus, or declining hearing, now should have a diagnostic MRI scan with contrast. Within 15 min, imaging answers the question as to whether an acoustic neuroma is present or not. We can now diagnose patients with absolutely normal hearing including speech discrimination. For such patients, a decision must be made as to whether to continue conservative management with serial observation, to consider surgical removal by one of several routes, or to perform radiosurgery [2, 3, 6, 7, 14–16, 21, 25, 26, 28–30,36–38].

Despite significant improvements in results of microsurgical excision at centers of excellence, residual patient morbidity has not been eliminated [15, 33, 41, 42]. The risks of craniotomy and tumor removal still include death, cerebrospinal fluid (CSF) leak, meningitis, postoperative hemorrhage, and cerebellar dysfunction, in addition to substantial risks of both temporary and permanent cranial nerve deficits. Large tumors with brainstem mass effect usually require surgical removal or debulking. Some elderly patients with hydrocephalus may best be managed by CSF diversion. However, the number of patients who are diagnosed late in the course of their disease will decline steadily.

During the past 10 years, the long-term safety and efficacy of stereotactic radiosurgery has established this surgical technique as an important alternative to microsurgery [10–12, 22–28]. At the University of Pittsburgh, approximately 120 patients each year undergo intervention for their acoustic tumors; 65% choose radiosurgery and 35% choose one of several microsurgical approaches. Stereotactic radiosurgery generally is reserved for patients with average extra-canalicular tumor diameters of <35 mm. Larger tumors routinely undergo microsurgical resection. Radiosurgery is performed subsequently for residual tumor as needed. Worldwide, an increasing shift of patients from microsurgical to radiosurgical alternatives has been observed. More than 1,800 patients have been treated by Gamma knife technique in the United States alone. In 1997, 500 patients had acoustic tumor radiosurgery at 27 Gamma knife sites. Since only 2000 patients each year are diagnosed in the United States, this represents an increasing percentage of patients. This expansion of the role of radiosurgery is based on more than 28 years of radiosurgery experience [9, 11, 12, 22–28, 31].

Conventional fractionated radiation therapy has been used for patients with residual acoustic neuromas that are observed to grow [40]. Such wide field multifraction techniques have been used sparingly, and the long-term tumor control rates as well as risk/complication rates are poorly defined. In 1969, under the direction of Lars Leksell, Georg Norén began a cautious program to use the newly created cobalt-60 Gamma knife for the management of acoustic neuromas [27]. The initial efforts were hampered by lack of ade-quate imaging (most patients required air encephalography for tumor defini-tion), and the absence of advanced dose-planning systems that would facilitate conformal single-fraction radiation of the tumor volume. Despite these initial difficulties, Norén's pioneering work laid the groundwork for other centers to expand upon the role and technique of radiosurgery for acoustic neuromas. In 1997, he presented a 20-year analysis of results at the American Association of Neurological Surgeons Meeting in Denver. Long-term tumor growth control was observed in more than 90% of patients. The development of high-reso-lution computed tomography (CT) imaging compatible with the evolving ste-reotactic head frame technology and the installation of the first North American 201 source Gamma knife allowed us to begin acoustic neuroma radiosurgery in 1987 [22–24, 28, 31]. The second patient treated at our center had an acoustic neuroma defined by CT imaging. During the course of our 10-year experience, we have sought to fulfill three initial goals. First, we wished to establish the role, risks and benefit of stereotactic radiosurgery in a long-term perspective. Second, we wished to take advantage of the steady advance-ments in imaging and computer dose-planning technology [9, 20]. Third, we wanted to be able to compare outcomes after radiosurgery to outcomes after microsurgery and to assess cost-effectiveness of such strategies. Pollock et al.

Table 1. Skull base schwannoma Gamma knife experience, center for Image-Guided Neurosurgery, University of Pittsburgh, 1987–1997

Schwannoma type	n
Acoustic (vestibular)	402
Trigeminal	12
Jugular foramen	15
Oculomotor	1
Total	430

[31] performed a matched cohort analysis of microsurgical and radiosurgical outcomes at our center. This study defined the long-term risk and benefit as well as patient outcomes and cost-effectiveness of radiosurgery when it is performed for tumors eligible for microsurgical excision at a center with extensive experience.

The University of Pittsburgh Experience

Since the introduction of the stereotactic radiosurgical technique with the gamma knife at the University of Pittsburgh in August 1987, a steady increase in case load of patients has been encountered. During this 10-year interval, 402 patients underwent Gamma knife radiosurgery for newly diagnosed, residual, or recurrent acoustic neuromas (table 1). Patient selection at our center often reflects a mutual decision between us, referring doctors, and patients and their families. Slightly more than half of our referrals come from primary care, ear, nose and throat surgeons, or neurosurgeons. Almost half come directly from patients themselves. Acoustic neuroma patients represent some of the best educated patients in the health care environment. Because their tumors are usually slow growing and benign, increasingly patients research surgical, conservative, and radiosurgical alternatives before making a decision. They often consult multiple physicians, obtain medical references, join patient support groups, and access Internet resources. It is our practice to provide consultative recommendations to patients and to referring doctors relative to the Gamma knife radiosurgical option. Patients are seen either immediately preceding their scheduled procedure or in advance in order to review all therapeutic options and the risks and benefits of the anticipated procedure. Informed consent is given by both the responsible neurological surgeon and the radiation oncologist in this multidisciplinary procedure.

A total of 402 patients (204 female, 198 male) underwent Gamma knife radiosurgery during a 10-year interval beginning in August 1987. The youngest patient was 14 and the oldest was 84, with a mean age of 57 years. The tumor was located on the right side in 204 patients and on the left side in 198 patients. Twenty-four patients had bilateral acoustic neuromas in the context of neurofibromatosis type II and 7 additional patients had unilateral tumors in the context of neurofibromatosis type II. Gross total resection had been performed in 26 patients (6.5%) and subtotal resection had been performed in 71 patients (17.7%), 18 patients having two or more surgical resections prior to Gamma knife radiosurgery. Preoperative hearing loss was present in 377 patients (94%), imbalance or ataxia was present in 214 (53%), tinnitus in 161 (41%), and dysphagia in 14 (3.5%). Preradiosurgical House Brackmann grade facial strength was described as grade I in 321 patients (80%), grade II in 27 patients (6.7%), grade III in 19 patients (44.7%), grade IV in 10 patients (2.5%), grade V in 7 patients (1.7%) and grade VI in 17 patients (4.2%). Trigeminal sensory loss was detected in 80 patients (19.9%). Seventy-eight patients had other neurological deficits. Fifty-two patients (12.9%) had a Karnofsky rating of 100, 302 patients (75%) had a Karnofsky rating of 90, and 48 patients had a Karnofsky rating of 70 or less. The Gardner-Robertson classification system was used to assess hearing in all patients [13]. Sixty-six patients (16.4%) had grade I hearing, 43 patients (10.7%) had grade II, 98 (24.4%) had grade III, 14 (3.5%) had grade IV, and 171 (42.5%) were deaf (grade V).

An average of 5.15 isocenters were used to treat the tumor with total coverage of the tumor possible in 396 patients (98.5%). A 4-mm beam diameter was used in 480 isocenters, 8 mm in 856 isocenters, 14 mm in 574 isocenters, and 18 mm in 172 isocenters. The 50% isodose or greater was used in 97.6% of patients.

Technical Evolution

During the past 10 years, our technique has steadily evolved [9]. The first change was in the imaging tool used to define tumor. Initially, CT imaging was used with iodinated contrast. Bone windows also helped to provide recognition of the intracanalicular portion of the tumor. With the demonstration of the accuracy and reliability of MRI, coupled with the recognition of its greatly improved contrast and spatial resolution, in the early 1990s we switched to MRI for virtually all patients [20]. Only those patients with residual metallic surgical clips or other reasons that preclude MRI scan undergo intraoperative CT scan-based planning. Our current technique for MRI is to perform a high-resolution volume acquisition with 1-mm slices through the tumor in the axial

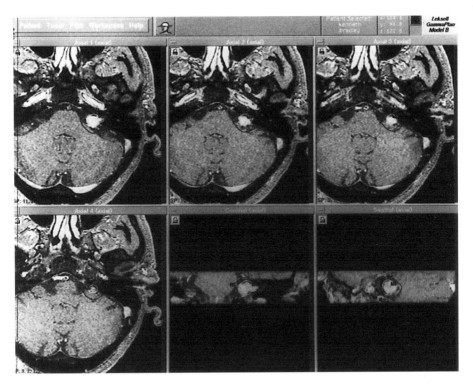

Fig. 1. Gamma plan dose planning for acoustic neuroma using axial 1-mm contrast-enhanced MRI with reformatted images in coronal and sagittal planes.

plane. These images are then reformatted into both coronal and sagittal planes using the current dose-planning system (fig. 1).

The second major evolution was to change from the original KULA dose-planning system (Elekta Instruments, Stockholm, Sweden), which was dependent upon slow calculations and generated isodose curves that could be overlaid over the images. In the 1990s, we started to use high-speed Hewlett-Packard computers. During the interval, we also developed and tested our own dose-planning system using a Silicone Graphics workstation computer. The 1997 planning system is Version 4.0 Gamma Plan (Elekta Instruments) which allows rapid calculation of multiple isocenters. With our current level of experience, computer dose planning often can be performed in less than 1 h depending in part on how quickly the images can be transferred via our hospital fiberoptic ethernet from the imaging site to our dose-planning center. Newer versions of Gamma Plan include a modification of the 'inverse

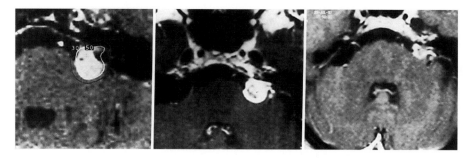

Fig. 2. Imaging response of a unilateral acoustic neuroma. Preoperative (left), 6 months postoperative (middle), and 6-year follow-up MRI (right).

algorithm' which enhances the computer generation of an idealized dose plan using parameters defined by the operating surgeon and physicist. In general, multiple isocenter Gamma knife radiosurgery (average five 'shots' per patient) is used to confine the irregular spherical geometry of the tumor. The final plan confines the desired isodose to the tumor margins identified by high-resolution imaging.

The third evolution in our technique has to do with gradual reduction in the marginal and maximal tumor doses based on continual evaluation of the results from our center as well as from other sites (fig. 2). Currently, doses at the margin using the 50% isodose are usually between 12 and 14 Gy. Doses are varied dependent upon tumor volume, hearing status and existing neurological function. Tumor dose reduction to date has not been associated with reduction in tumor growth control (fig. 3).

The fourth evolution in our technique has been switching of patients to treatment with the Model B Unit Gamma knife which was installed in our center in 1996. This first North American B Unit device has proven to be very efficient. Since beam channel blocking is usually not necessary because of the reconfiguration of the sources into a circular array, and because of the more newly loaded cobalt energy sources, initial treatment times were comparatively low. To date, we have not noticed any specific dose-rate effect in terms of risks or benefits to patients and to dose levels of 12–14 Gy at the margin. It is possible that newly loaded units with dose rates of up to 400 cGy/min may be associated with a greater adverse risk of complications (cranial nerve dysfunction) if doses ≥15 Gy are used. When the current marginal doses are used, a dose-rate effect appears undetectable.

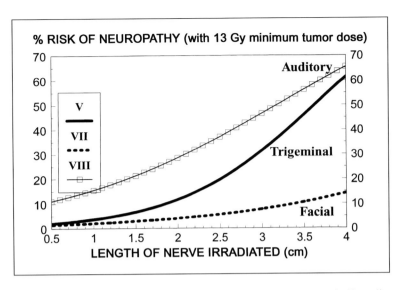

% RISK OF NEUROPATHY (with 13 Gy minimum tumor dose)

Fig. 3. The risk of cranial nerve dysfunction after Gamma knife radiosurgery related to dose delivered to the tumor margin.

Tumor Control after Gamma Knife Radiosurgery

Follow-up recommendations include serial MRI scans at 6, 12, 24, 48 and 96 months. For patients who have preoperative hearing, audiograms are requested at the same time as long as hearing can be detected. The MRI scan is done with and without contrast as well as with proton density images to assess signal alterations in adjacent brain. Tumor measurements are made on all follow-up scans as described by Linskey et al. [22]. The anterior-posterior dimensions (parallel to the petrous bone), the left-right dimension perpendicular to the IAC (extracanalicular dimension), and the superior-inferior dimension (coronal plane) are measured and recorded. Significant volume reduction was considered when the total tumor volume was 25% or less. Imaging studies also assessed the risk of peritumoral brainstem or cerebellar effects using long-TR imaging. Follow-up varied from 3 months to 10 years with a mean follow-up of 3 years. In our initial reviews, tumor volumes decreased in 30% of patients, remained unchanged in 63.5%, and subsequently increased in 6.5%, including 8 patients (2%) who had delayed surgical resection (table 2). Flickinger et al. [10] evaluated the evolution of technique for acoustic neuroma management in our center, analyzing the results of 273 patients with unilateral tumors. However, in the long-term analysis of patients managed between 1987 and 1992, over 60% of patients had reduction in tumor volume. One hundred

Table 2. Acoustic neuroma management in 402 patients'
imaging response after radiosurgery

25% tumor volume change	% of patients
Decreased[a]	30
No change	63.5
Increased[b, c]	6.5

[a] 5-year follow-up 62% had volume reduction.
[b] Three patients (0.7%) developed mass effect.
[c] Eight patients (2%) had delayed surgical resection.

Table 3. Acoustic neuroma management in 402 patients'
cranial nerve function after radiosurgery (% of patients with
preservation of function)

Cranial nerve	CT-based planning	MRI-based planning
Trigeminal	64	92
Facial	72	92
Auditory	39	68

and eighteen patients underwent CT-based planning between 1987 and 1991, and 155 patients had MRI-based planning between 1991 and 1994 (table 3). During this interval, tumor dose also was reduced from an average of 17 to 14 Gy at the margin and the mean number of isocenters used per tumor increased from 3.4 to 5.8. Tumor volumes were slightly lower in the MRI series (2.7 vs. 3.5 cm^3 in the CT series). The actuarial 7-year clinical tumor control rate (i.e. no requirement for further surgical intervention) was 96.4 ± 2.3%. The radiographic tumor control rate was 91 ± 3.4% which was unchanged between CT and MRI groups.

Loss of central contrast enhancement frequently occurs between 6 and 12 months after radiosurgery. This appears to correlate best with the high radiation isodoses. The effect is often transient since contrast reuptake re-develops within another 6 months. Loss of contrast is positively predictive of delayed shrinkage of the tumor over the ensuing 2 years. Peritumoral brainstem edema has become increasingly rare and has been observed in the interval from 1993 to 1997. This comparative reduction likely is associated with our ability to do conformal planning using MRI, the increased number of smaller beam diameter isocenters, and reduced dosage. The development of hydrocephalus was a rare event in our series, occurring in 8 patients, 3 of whom had hydrocephalus prior to Gamma knife radiosurgery. In addition, several other patients already had shunts placed prior to radiosurgery.

Failed Microsurgery and the Role of Radiosurgery

In a collaborative assessment between our center and that of the Mayo Clinic under the direction of Drs. Robert Coffey and Bruce Pollock, we evaluated the results of stereotactic radiosurgery in 72 consecutive patients with prior surgical resection referred to our respective centers during a 9-year interval. All patients had a minimum of 1-year follow-up. All patients had tumors either recurrent after prior total resection (34%) or persistent after subtotal resection (66%). Twenty-eight patients (38%) had undergone multiple resections. Forty-one patients (55%) had significant impairment of facial nerve function (House Brackmann grade III–VI) after their microsurgical procedure. Fifty percent had trigeminal loss and 96% had unserviceable hearing (pure tone average >50 dB and speech discrimination $<50\%$). Median follow-up after radiosurgery was 40 months (range 12–97). Tumor control after radiosurgery was achieved in 69 patients (93%). Six patients required additional resection despite radiosurgery at a median of 32 months afterwards. One patient had repeat radiosurgery for tumor progression outside the previously irradiated volume. Nine of 45 patients (20%) with grade I–III facial function prior to radiosurgery developed increased facial weakness after radiosurgery, and 10 patients (14%) developed new trigeminal symptoms. Our experience in patients with prior microsurgery suggests that the risk of radiosurgery complications may be greater in patients who had microsurgery. Because of the large number of patients we have seen whose tumors progressed despite initial microsurgery, we believe that the true risk of progression of removal of acoustic neuromas in this country may be underreported by the surgical literature [15, 32, 34]. Although certain studies have recently estimated this risk to be as low as 1%, we believe that the actual risk in the United States is between 5 and 10% at 10 years [41, 42].

Failed Radiosurgery and the Role of Microsurgery

Because of the controversy relating to whether microsurgery after failed radiosurgery is more difficult, we also reviewed our 10-year combined experience at Pittsburgh and the Mayo Clinic. During this 10-year interval,

523 patients at both centers underwent Gamma knife radiosurgery. Eleven patients with unilateral tumors underwent delayed microsurgery at 6–49 months after radiosurgery. Six of these patients had undergone microsurgery prior to radiosurgery. Delayed microsurgery consisted of suboccipital removal in 9 patients, translabyrinthine removal in 1 and a combined approach in 1 patient. Gross total tumor removal was possible in only 6 patients. The tumor's consistency was described as cystic in 2, adherent to adjacent structures in 5 (3 had undergone prior microsurgery), fibrous in 2, and normal in 5. Postoperative complications included hydrocephalus in 5, meningitis in 2, CSF leak in 2, and facial palsy in 2 patients. Failed radiosurgery was rare at our two institutions, occurring in <2% of patients and always before 5 years had elapsed after radiosurgery. Failed radiosurgery was more common in patients who had already failed microsurgery. Clinical worsening was frequent when microsurgery was performed after radiosurgery and patient outcomes usually deteriorated. The indications for microsurgery after radiosurgery must include progressive and sustained growth with new neurological symptoms. Since 1–2% of patients have slight tumor growth before subsequent growth arrest, the time needed for delayed microsurgical resection should be reviewed with the surgeon who performed radiosurgery.

Neurofibromatosis, Type II

We performed radiosurgery during a 10-year interval on 24 patients with bilateral tumors (NF-II). Most patients had undergone prior surgical removal of at least one of the tumors. Most patients had significant preoperative hearing loss. Tumor growth control was achieved in 70% of these patients. However, because of the tendency of the NF-II tumor to engulf the hearing nerve, which may still remain functional, we normally reserve radiosurgery for those patients with small- to medium-sized tumors that are shown to progress by MRI and are associated with progressive hearing deterioration. The decision to treat patients with NF-II is difficult and becomes increasingly so when the tumor progresses in an only hearing ear. Although radiosurgery can preserve functional hearing in 30–50% of NF-II patients, this result is dependent on multiple isocenter dose planning and tumor marginal doses of approximately 12 Gy. We believe that stereotactic radiosurgery represents one of the multiple modalities often needed during the course of management of patients with NF-II and should be considered as a first-line option for patients with small- to medium-sized tumors who begin to have declining hearing and preserved contralateral hearing.

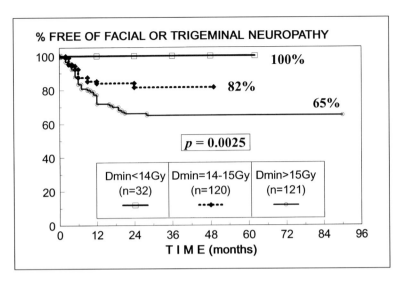

% FREE OF FACIAL OR TRIGEMINAL NEUROPATHY

Fig. 4. Improvement in cranial nerve outcomes related to changes in tumor margin dose.

Cranial Nerve Function

In our studies, we have identified a number of factors which have improved cranial nerve preservation rates. In comparison to our earlier reports, MRI dose planning, increasing the number of isocenters, reducing the beam diameters used, and reducing the dose to an average marginal dose of 12–14 Gy have resulted in steady risk reduction for trigeminal, facial and auditory nerve preservation (fig. 4). At the present time, using MRI dose planning, facial nerve function at the preoperative level can be preserved between 96 and 100% of patients. Trigeminal nerve dysfunction occurs in <10% of patients with extracanalicular tumors near the trigeminal nerve and in virtually no patients with intracanalicular tumors. In contrast to microsurgery, hearing nerve preservation is always 100% immediately after radiosurgery. Hearing preservation can be maintained at the preoperative level in 50–70% of patients, with an additional 15–20% of patients having preserved but worsened hearing, and 10–15% of patients becoming deaf. Relatively few patients develop new and persistent imbalance or vestibular complaints. Between 5 and 18 months after the procedure, approximately 15% of patients will develop a transient or temporary increase in imbalance or gait instability. In patients with tinnitus, approximately one-third remain unchanged, one-third show some improvement, and one-third will show temporary exacerbation between 5 and 18 months followed by a return to baseline. Only 1 patient with disabling residual tinnitus has been seen in our experience in more than 400 patients.

Patient Outcomes

In a retrospective matched cohort series comparing microsurgery and radiosurgical patients treated at our institution, Pollock et al. [31] identified a high patient satisfaction rate, and an overall tendency towards better quality of life in patients treated by radiosurgery in comparison to microsurgery. Radiosurgery patients are able to leave the hospital within 24 h, return to their usual occupation in 2 or 3 days, and maintain their preoperative level of employment status in >98% of cases. Recently, Kondziolka et al. [20] completed a long-term outcome study from the patient's perspective. In this study, patient expectations, levels of satisfaction, and complication rates were measured. One hundred and sixty-two patients treated in the first 5-year interval of our experience were reviewed between 5 and 10 years after radiosurgery. Forty-two patients (26%) had already had prior microsurgery. As in our entire series, the clinical tumor control rate (no need for additional resection) was 98%. For patients with at least 7 years of follow-up, the tumor control rate was 94%. At this time, 62% of patients had tumors that were smaller, 53 (33%) had tumors that remained unchanged in size, and 9 (6%) were larger. Ten patients had died of unrelated causes during the follow-up interval.

A patient survey was returned by 115 patients. Patient activity levels remained unchanged in 68%, increased in 8% and gradually decreased in 24%. Gamma knife radiosurgery met patient expectations in 106 (92%) of patients, including 97% of the prior surgery group and 91% in the no-resection group. Complications were described by 36 patients (31%), 55% of which resolved. These included balance problems (n = 7), facial twitching (n = 6), persistent facial weakness (n = 4), tinnitus (n = 3), hydrocephalus (n = 3), and headache (n = 2). Radiosurgery was believed successful in 30 patients who had undergone prior surgery and 96% of those who had not had prior resection. When asked if they would recommend this treatment for a family member or friend, 95% replied 'yes'. This study provided additional long-term evidence of tumor control, and contrasts greatly with the outcome studies reported by the Acoustic Neuroma Association [41, 42].

Evolution and Revolution

Inherent training bias continues to inhibit rational discourse relative to the selection of treatment options for patients with acoustic neuromas. Tumors are now being detected much earlier, even in patients with minimal hearing loss, mild imbalance, or recent onset of tinnitus. Although serial observation remains a reasonable option for elderly patients with potentially slow growing tumors [6, 38], we believe that a revolution in thinking is in order. We believe that gamma knife stereotactic radiosurgery will become the primary management option for patients with small- to moderate-sized growing acoustic tumors. In order for this to occur, patients should undergo confirmatory neurodiagnostic testing early when the tumor is first suspected. Patients who are willing to accept tumor growth control rather than tumor removal can reasonably select radiosurgery as the first-line treatment option. Gamma knife radiosurgery is a multidisciplinary surgical field using the talents of physicists, radiation oncologists, radiologists and neurosurgeons. Neurosurgeons must maintain a prominent role in patient selection and service delivery, especially since complication detection and management is usually left to the neurosurgeon. Training programs must provide an in-depth experience. In the future, both radiosurgical and microsurgical treatment options should be explained to all patients and be available in centers of excellence where physicians have been jointly trained in both techniques and have extensive experience in both treatment options [25]. Our experience indicates that radiosurgery is increasingly likely to be demanded by patients and their insurance carriers. During our 10 year experience, tumor growth control, symptom control, cranial nerve preservation, improved outcomes, and reduced costs have been demonstrated.

References

1   Beatty CW, Ebersold MJ, Harner SG: Residual and recurrent acoustic neuromas. Laryngoscope 1987;97:1168–1171.
2   Bederson JB, Von Ammon K, Wichmann WW, Yasargil MG: Conservative treatment of patients with acoustic neuromas. Neurosurgery 1991;28:646–651.
3   Cerullo LJ, Grutusch JF, Heiferman K, Osterdock R: The preservation of hearing and facial nerve function in a conservative series of unilateral vestibular nerve schwannoma surgical patients. Surg Neurol 1993;39:485–493.
4   Charabi S, Thomsen J, Mantoni M, et al: Acoustic neuroma (vestibular schwannoma): Growth and surgical and nonsurgical consequences of the wait-and-see policy. Otolaryngol Head Neck Surg 1995;113:5–14.
5   Cox DR: Regression models and life tables. J R Stat Soc B 1972;34:187–220.
6   Deen HG, Ebersold MJ, Harner SG, et al: Conservative management of acoustic neuroma: An outcome study. Neurosurgery 1996;39:260–266.
7   Fischer G, Fischer C, Remond J: Hearing results of the retrosigmoid approach to acoustic neuroma. J Neurosurg 1992;76:910–917.
8   Flickinger JC: An integrated logistic formula for prediction of complications from radiosurgery. Int J Radiat Oncol Biol Phys 1989;17:879–885.
9   Flickinger JC, Kondziolka D, Pollock BE, Lunsford LD: Evolution in technique for vestibular schwannoma radiosurgery and effect on outcome. Int J Radiat Oncol Biol Phys 1996;36:275–280.
10   Flickinger JC, Lunsford LD, Linskey ME, et al: Gamma knife radiosurgery for acoustic tumors: Multivariate analysis of four-year results. Radiother Oncol 1993;27:91–98.
11   Foote Rl, Coffey RJ, Swanson JW, et al: Stereotactic radiosurgery using the gamma knife for acoustic tumors. Int J Radiat Oncol Biol Phys 1995;32:1153–11160.

12 Forster DMC, Kemeny AA, Pathak A, Wlaton L: Radiosurgery: A minimally interventional alternative to microsurgery in the management of acoustic neuroma. Br J Neurosurg 1996;10:169–174.

13 Gardner G, Robertson JH: Hearing preservation in unilateral acoustic neuroma surgery. Ann Otol Laryngol 1988;97:55–66.

14 Glasscock ME III, Hays JW, Minor LB, et al: Preservation of hearing in surgery for acoustic neuroma. J Neurosurg 1993;78:864–870.

15 Gormley WB, Sekhar LN, Wright DC, Kamerer D, Schessel D: Acoustic neurons: Results of current surgical management. Neurosurgery 1997;41:50–60.

16 Haines SJ, Levine SC: Intracanicular acoustic neuroma: Early surgery for preservation of hearing. J Neurosurg 1993;79:515–520.

17 House JW, Brackmann DE: Facial nerve grading system. Otolaryngol Head Neck Surg 1985;93:146–147.

18 Hudgins WR: Patients' attitude about outcomes and the role of gamma knife radiosurgery in treatment of vestibular schwannomas. Neurosurgery 1994;34:459–465.

19 Kaplan EL, Meier P: Nonparametric estimation and incomplete observations. J Am Stat Assoc 1958;53:457–481.

20 Kondziolka D, Dempsey PK, Lunsford LD, Kestle JR, et al: A comparison between MRI and CT for stereotactic coordinate determination. Neurosurgery 1992;30:402–406.

21 Laasonen EM, Troupp H: Volume growth rate of acoustic neurinomas. Neuroradiology 1986;28:203–207.

22 Linskey ME, Flickinger JC, Lunsford LD: The relationship of cranial nerve length to the development of delayed facial and trigeminal neuropathies after stereotactic radiosurgery for acoustic tumors. Int J Radiat Oncol Biol Phys 1993;25:227–234.

23 Linskey ME, Lunsford LD, Flickinger JC: Stereotactic radiosurgery for acoustic neuromas: Early experience. Neurosurgery 1990;26:933–938.

24 Linskey ME, Lunsford LD, Flickinger JC: Tumor control stereotactic radiosurgery in neurofibromatosis patients with bilateral acoustic tumors. Neurosurgery 1992;31:829–839.

25 Mendenhall WM, Friedman WA, Bova FJ: Linear accelerator-based stereotactic radiosurgery for acoustic schwannomas. Int J Radiat Oncol Biol Phys 1994;28:803–810.

26 Norén G, Hirsch A, Mosskin M: Long-term efficacy of gamma knife radiosurgery in vestibular schwannomas. Acta Neurochir 1993;62(suppl):122–164.

27 Norén G, Griettz D, Hirsch A, Lax I: Gamma knife surgery in acoustic tumors. Acta Neurochir 1993;58(suppl):104–107.

28 Ogunrinde OK, Lunsford LD, Flickinger JC, Kondziolka D: Stereotactic radiosurgery for acoustic nerve tumors in patients with useful preoperative hearing: At 2-year follow-up. J Neurosurg 1994;80:1011–1017.

29 Ojemann RG: Management of acoustic neuromas (vestibular schwannomas). Clin Neurosurg 1993;40:498–535.

30 Peto R, Pike MC, Armitage P, et al: Design and analysis of randomized clinical trials requiring prolonged observation of each patient. Br J Cancer 1977;35:1–39.

31 Pollock BE, Lunsford LD, Kondziolka D, et al: Outcome analysis of acoustic neuroma management: A comparison of microsurgery and radiosurgery. Neurosurgery 1995;36:215–229.

32 Post KD, Eisenberg MB, Cataland PJ: Hearing preservation in vestibular schwannoma surgery: What factors in future outcome? J Neurosurg 1995;83:191–196.

33 Samii M, Mathies C: Intracanalicular acoustic neurinomas. Neurosurgery 1991;29:189–199.

34 Samii M, Matthies C: Management of 1,000 vestibular schwannomas (acoustic neuromas): Surgical management and results with an emphasis on complications and how to avoid them. Neurosurgery 1997;40:11–23.

35 Schessel DA, Nedzelski JM, Kassel EE, Rowed DW: Recurrence rates of acoustic neuroma in hearing preservation surgery. Am J Otol 1992;13:233–235.

36 Silverstein H, McDaniel A, Norrel H, Wazen J: Conservative management of acoustic neuroma in the elderly patient. Laryngoscope 1985;95:766–770.

37 Slattery WH, Brackmann DE: Results of surgery following stereotactic irradiation for acoustic neuromas. Am J Otol 1995;16:315–319.

38   Strasnick B, Glasscock ME III, Haynes D, et al: The natural history of untreated acoustic neuromas. Laryngoscope 1994;104:1115–1119.
39   Umezu H, Aiba T, Tsuchida S, Seki Y: Early and late postoperative hearing preservation in patients with acoustic neuromas. Neurosurgery 1996;39:267–272.
40   Wallner KE, Sheline GE, Pitts LH, et al: Efficacy of irradiation for incompletely excised acoustic neurolemmomas. J Neurosurg 1987;67:858–863.
41   Wiegand DA, Fickel V: Acoustic neuroma: The patient's perspective: Subjective assessment of symptoms, diagnosis, therapy, and outcome in 541 patients. Laryngoscope 1989;99:179–187.
42   Wiegand DA, Ojemann RG, Fickel V: Surgical treatment of acoustic neuroma (vestibular schwannoma) in the United States: Report from the acoustic neuroma registry. Laryngoscope 1996; 106:58–66.
43   Weit RJ, Zappia JJ, Hecht CS, et al: Conservative management of patients with small acoustic tumors. Laryngoscope 1995;105:795–800.

L. Dade Lunsford, MD, University of Pittsburgh Medical Center, Suite B-400,
200 Lothrop Street, Pittsburgh, PA 15213 (USA)
Tel. (412) 647 6781, Fax (412) 647 0989

Lunsford LD, Kondziolka D, Flickinger JC (eds): Gamma Knife Brain Surgery.
Prog Neurol Surg. Basel, Karger, 1998, vol 14, pp 104–113

....

Stereotactic Radiosurgery of Meningiomas

Douglas Kondziolka, L. Dade Lunsford, John C. Flickinger

Departments of Neurological Surgery and Radiation Oncology and
The Center for Image-Guided Neurosurgery, University of Pittsburgh, Pa., USA

Surgical resection is the preferred treatment of meningiomas when elimination of both the tumor mass and the neoplastic dura can be achieved. In this setting, long-term disease-free survival is expected [17]. In some patients such complete surgical resection may be associated with the undesirable development of new neurological deficits. In certain brain locations, particularly around the skull base or sagittal sinus, total resection is not feasible and the neoplastic dural base may either be only coagulated or left intact. In those patients where surgical cure is not anticipated or the risks of new neurologic deficits seem excessive, Gamma knife radiosurgery represents an important therapeutic option. In 1991 we reported the first evaluation of the use of radiosurgical techniques for meningiomas [11]. Since that time, meningioma radiosurgery has been performed frequently at centers worldwide. Concomitant with improvements in microsurgical resection, great improvements in radiosurgery dose planning, computer workstations, and our understanding of the biology and dose-volume relationships of radiosurgery have improved results. Over the last 10 years, we have acquired a much more intensive understanding of normal vascular, cranial nerve and parenchymal effects, and the response of the tumor itself.

University of Pittsburgh Experience

Over a 10-year interval, 314 patients with cranial meningiomas had stereotactic radiosurgery at our center. Two hundred thirteen patients were female (68%) and 101 were male. The mean patient age was 57 years (range 12–86). The average duration of symptoms before radiosurgery was 2 years. Seventy-

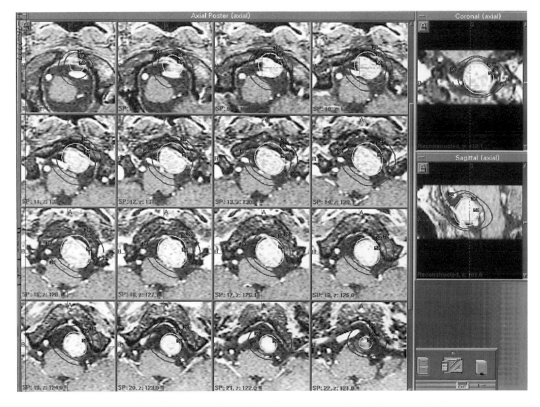

Fig. 1. Gamma knife radiosurgery dose plan for a woman with a clivus meningioma anterior to the medulla. Coronal and sagittal images are shown on the right. One 14-mm and three 8-mm isocenters were used to deliver a 50% margin dose of 14 Gy.

five percent of patients had a neurological deficit before radiosurgery, 25% complained of headaches, and 10% had sustained at least one seizure. A subtotal tumor resection had been performed in 148 patients (47%) and 27 additional patients had already undergone 'gross total' resection (9%). Thirty patients had received prior external beam radiation therapy (9.6%). One patient had undergone prior meningioma embolization. Eighty percent of these meningiomas were located at the cranial base. The most common sites were the cavernous sinus and petroclival regions. Tumors arising from the cerebral hemisphere convexities were encountered less frequently.

Multiple isocenter radiosurgery was used in the vast majority of patients (range 1–14) (fig. 1). The average number of isocenters per patient was 6.1. Radiosurgery was targeted to the enhancing tumor margins. We did not 'chase' dural tails beyond an area of obvious dural widening by tumor. The mean tumor volume was 4.7 ml (maximum 28.5). The 50% isodose line was used in 276 patients (88%). Lower isodoses included 20% (n = 1), 30% (n = 1), 35% (n = 1), 40% (n = 14) and 45% (n = 7). The mean dose delivered to the tumor margin was 15 Gy (maximum 34). The mean maximum tumor dose was 30.1 Gy (maximum 54).

Neurologic Symptoms after Radiosurgery

Most patients were unchanged during prolonged follow-up after radiosurgery. Four elderly patients had placement of a ventriculoperitoneal shunt after radiosurgery. Four patients developed new visual symptoms after radiosurgery. Two of these patients had reduced vision because of radiation-related optic chiasmal effects (one had temporary decrease in visual acuity with recovery and another had a persistent hemianopsia). One patient developed edema in the calcarine region. One additional patient suffered a decline in vision 5 months after radiosurgery for a radiation-induced meningioma of the tuberculum. Three patients developed new trigeminal symptoms and 4 suffered a transient hemiparesis. Fourteen patients had delayed deaths despite radiosurgery; however, only 1 patient died from tumor progression (this patient had a malignant meningioma with metastases).

Imaging Response

In 226 patients with minimum 6-month follow-up, 6-patients developed total resolution of their tumor (2%), 101 had a decrease in tumor volume (45%), 108 had no change in tumor volume (48%) and 11 had an increase (5%) (fig. 2). This 95% control rate is similar to the 94% rate identified after acoustic tumor radiosurgery [7], although a higher percentage of patients with tumor reduction was found in the acoustic series. Peritumoral imaging changes (suggestive of edema) developed in 21 patients (7%); most affected patients had no accompanying clinical symptoms.

Expectations after Meningioma Management

The classic 1957 paper by Simpson [17] provided data on meningioma recurrence. He reported a 9% recurrence rate for tumors completely resected along with their dural attachment, a 19% recurrence rate if the tumor was completely resected and the adjacent dura cauterized, and a 40% recurrence rate in patients

---

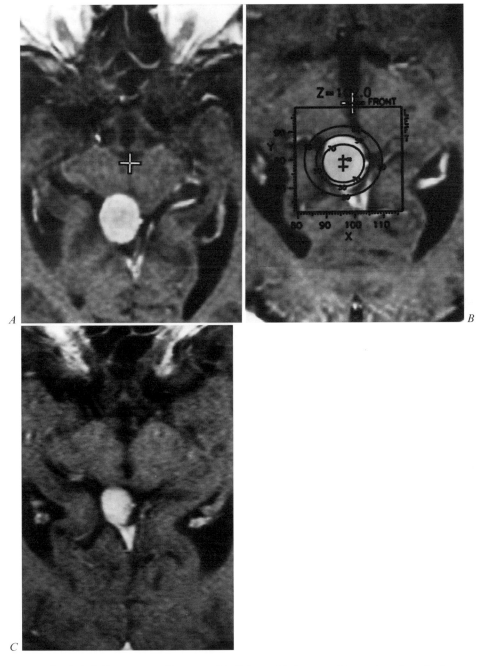

Fig. 2. Axial images at radiosurgery (*A, B*) in an elderly man with a tentorial meningioma. The 50% isodose line is at the tumor margin. 4.5 years later (*C*), the tumor was slightly smaller.

with partial resection. A recent report by Condra et al. [1] from the University of Florida showed that subtotal resection alone was associated with a poor long-term prognosis including a strong relation to disease-specific mortality. At 15 years, local control was 76% after 'total' excision and 87% after subtotal excision followed by tumor irradiation. Patients who had only subtotal resection had only a 30% local control rate. Since the tumor growth control rate after radiosurgery (at least with results between 5–10 years) is above 90%, radiosurgery has at least a longer term tumor control rate that equals total tumor resection including removal of the neoplastic dura. For any resection where this is not achieved, the long-term result (in terms of tumor control) will be worse than that achieved with radiosurgery. Thus, for a tumor that would be suitable for radiosurgery a priori, surgical resection should be performed if a cure is expected. For meningiomas along the convexity, falx, lateral sphenoid ridge, and anterior fossa floor, we believe a resection is an excellent first strategy. For all other meningiomas, radiosurgery may prove to be a better first option unless the tumor has caused progressive mass effect, intractable epilepsy, or some other problematic neurologic syndrome that requires a semi-urgent reduction in tumor volume. Larger tumors that are closely related to critical structures such as the optic chiasm are best managed with combined strategies of resection and radiosurgery or resection and radiation therapy.

Although some earlier series on meningioma radiation therapy (from the 1980s) did not show a convincing benefit to this therapy, more recent reports using modern imaging and current radiation techniques show a benefit, particularly for a tumor after subtotal resection [1, 10, 18].

Biologic Response of Meningiomas to Radiosurgery

Little is known regarding the specific histologic response of meningiomas to single-fraction irradiation. We suspect that tumor control and later regression is due to both meningioma cell injury (leading to apoptosis which may occur only at the time of attempted cell division) and to delayed vascular obliteration. Although loss of central contrast enhancement is more typical of acoustic tumors than of meningiomas, this finding does occur and often predicts later tumor regression. When we implanted human meningiomas into the subrenal capsule of athymic mice, we noted early regression in tumor volume and vascularity after maximum doses of 20 and 40 Gy (within a 3-month period). This response was similar to that identified in a similar experimental study using acoustic tumors [14].

It is clear that the meningioma response occurs slowly and progressively. Tumor regression may take up to several years following radiosurgery. Interestingly, the large normal blood vessels within meningiomas (such as those within the cavernous sinus) do not reduce in caliber following radiosurgery despite their receiving the higher central radiation dose [2, 13]. We have not observed that such vessels, even when narrowed by the tumor, develop progressive obliteration. In similar fashion, small perforating arteries that surround meningiomas do not appear to occlude since we have seen no patient develop a lacunar infarction or other imaging change following any skull base radiosurgery procedure [5].

Malignant Meningiomas

In our early experience, we used radiosurgery as the sole radiation modality for patients with small-volume malignant meningiomas. However, this proved to be an ineffective strategy by itself and all patients required eventual radiation therapy for tumor delayed growth. Since radiation therapy by itself also does not appear to be enough for such tumors, we use both radiation modalities for management [12]. We aggressively reevaluate meningioma specimens from outside medical centers and reconfirm the histologic diagnosis of a truly benign meningioma. Since atypical, angioblastic and malignant meningiomas all portend a worse prognosis, we often consider radiation therapy for treatment. On the other hand, hemangiopericytomas respond promptly and dramatically to radiosurgery [13].

Radiosurgery at the Skull Base

Eighty percent of our meningioma experience has centered on tumors at the skull base. The difficulties posed by tumors in these brain locations make total resection difficult and potential morbidity greater than in most supratentorial locations (fig. 3). Thus, if radiosurgery is to be an effective management for a skull base meningioma, it must provide long-term tumor growth control and be free of cranial nerve or brainstem morbidity [2] (fig. 4). Along with our skull base meningioma series, we have developed a large experience in acoustic tumor radiosurgery (over 400 cases), pituitary tumor radiosurgery (over 100 cases), and radiosurgery for other schwannomas and some malignant skull base tumors [13]. We have identified a consistent response of cranial nerves to radiosurgery regardless of tumor pathology. For most part, these nerves are tolerant of radiosurgery doses between 12 and 18 Gy but this response may vary depending on length of nerve irradiated, nerve ischemia, degree of compression by tumor, prior surgery or radiation therapy, and

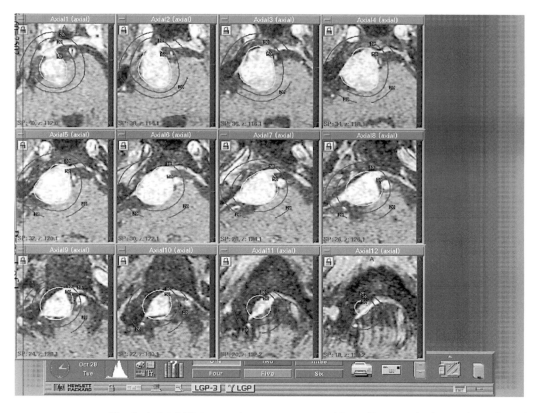

Fig. 3. Axial MR images in a young man with a clivus meningioma. A conformal radiosurgery volume was created that remained tight on the brainstem surface. Although a dural tail can be seen, the tumor did not enter the cavernous sinus.

possibly patient age [4, 6, 8, 15, 18]. Nevertheless, we expect motor nerve morbidity rates of less than 5% in most patients.

The special sensory nerves pose a more difficult problem. Although cochlear nerve function remains the most pressing issue in acoustic tumor radiosurgery, it appears to be of less importance in meningioma radiosurgery, even for tumors in the cerebellopontine angle. We have not observed deafness after management of a meningioma in this location or along the petrous ridge. This may be related to differences in nerve ischemia and compression between meningiomas and schwannomas. We would expect schwannomas to cause more nerve effects because of their intimate relation and evolution from the nerve. The optic nerve remains important in cavernous sinus or parasellar radiosurgery [17]. Since 2 of our early patients (1988) developed optic neuropathy at doses between 11 and 12.5 Gy, subsequently we have limited the optic chiasm or nerve dose to below

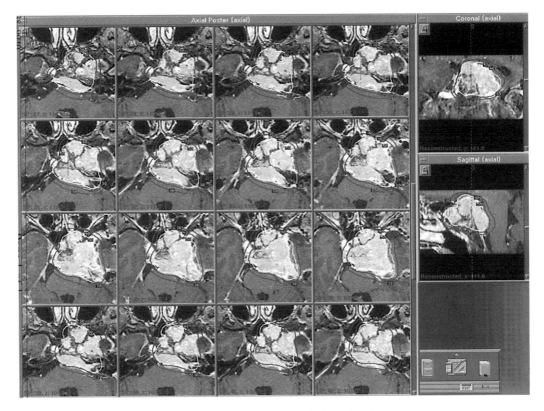

Fig. 4. Radiosurgery plan for a left cavernous sinus meningioma in a 51-year-old woman. She had prior resection of middle fossa tumor. A plan was made with five 18-mm and seven 14-mm isocenters to deliver a maximum dose of 22 Gy. Note the steep fall-off in radiation near the optic chiasm (upper right).

8 Gy utilizing advanced dose-planning methods, beam blocking, and smaller isocenters [9]. Since that time, no meningioma patient has developed an optic neuropathy. We have not observed hypopituitarism after radiosurgery for cavernous sinus or parasellar meningiomas.

Radiosurgery around the Venous Sinuses

We have pursued the role of radiosurgery for management of meningiomas of the sagittal and transverse sinus regions. Since an intimate relationship with the sinus often precludes total resection, we have managed many patients with residual or recurrent tumors in these locations, or in occasional older patients with small volume tumors as a primary treatment. In our initial experience, we did not identify any increased risk to bridging veins or the sinus. We did not observe an increased incidence of regional brain edema or venous infarction. However, a detailed evaluation of a large patient series will be necessary before a formal conclusion can be made regarding the tolerance of venous sinuses to radiosurgery.

Even if radiosurgery was to cause progressive sinus occlusion, we anticipate that this occurrence would be so slow as to be asymptomatic in most patients (similar to thrombosis induced by tumor growth). Most referred patients have evidence of regional encephalomalacia in the region of surgical resection. This area might provide a 'buffer zone' for radiosurgery since tissue damage already is present. We are currently evaluating results from 20 Gamma knife centers who have participated in a multicenter review of parasagittal meningioma radiosurgery. We hope to clarify the expectations of radiosurgery in this location and develop a new strategy for the difficult problem of the recurrent parasagittal meningioma.

Repeat Radiosurgery

Eight patients in this series underwent an additional radiosurgery procedure for management of tumor growth outside the initial radiosurgery volume. The most frequent site for this was along the falx and superior sagittal sinus, where an initially normal-appearing dura later developed marked expansion consistent with meningioma growth. Since there likely exists a region of neoplastic dura from which a meningioma can arise, such an occurrence is not unexpected.

Summary

Meningioma radiosurgery is one of the most challenging aspects of stereotactic surgery. Patient-specific decisions need to be made regarding the imaging-defined tumor anatomy, radiation dose, long-term expectations, and relative risks and benefits of other therapies. Our 10-year experience has documented great resistance for most cranial nerves that surround meningiomas, and acceptable tolerance of surrounding brain or brainstem parenchyma and blood vessels. We believe that the aggressive skull base surgeries of the last decade will be replaced by the judicious combined use of microsurgical and radiosurgical techniques to improve overall patient outcomes.

References

1 Condra K, Buatti J, Mendenhall WM, et al: Benign meningiomas: Primary treatment selection affects survival. Int J Radiat Oncol Biol Phys 1997;39:427–436.
2 Duma CM, Lunsford LD, Kondziolka D, et al: Stereotactic radiosurgery of cavernous sinus meningiomas as an addition or alternative to microsurgery. Neurosurgery 1993;32:699–705.
3 Engenhart R, Kimmig B, Hover K, et al: Stereotactic single high-dose radiation therapy for benign intracranial meningiomas. Int J Radiat Oncol Biol Phys 1990;19:1021–1026.
4 Flickinger JC: An integrated logistic formula for prediction of complications from radiosurgery. Int J Radiat Oncol Biol Phys 1989;17:879–885.
5 Flickinger JC, Kondziolka D, Kalend AM, Maitz AH, Lunsford LD: Radiosurgery-related imaging changes in surrounding brain: Multivariate analysis and model evaluation; in Kondziolka D (ed): Radiosurgery 1995. Radiosurgery. Basel, Karger, 1996, vol 1, pp 229–236.
6 Flickinger JC, Kondziolka D, Lunsford LD: Dose and diameter relationships for facial, trigeminal, and acoustic neuropathies following acoustic neuroma radiosurgery. Radiother Oncol 1996;41:215–219.
7 Flickinger JC, Kondziolka D, Pollock BE, et al: Evolution of technique for vestibular schwannoma radiosurgery and effect on outcome. Int J Radiat Oncol Biol Phys 1996;36:275–280.
8 Flickinger JC, Lunsford LD, Kondziolka D: Dose prescription and dose-volume effects in radiosurgery. Neurosurg Clin North Am 1992;3:51–59.
9 Flickinger JC, Maitz A, Kalend A, et al: Treatment volume shaping with selective beam blocking using the Leksell gamma unit. Int J Radiat Oncol Biol Phys 1990;19:783–789.
10 Glaholm J, Bloom H, Crow JH: The role of radiotherapy in the management of intracranial meningiomas: The Royal Marsden Hospital experience with 186 patients. Int J Radiat Oncol Biol Phys 1990;18:755–761.
11 Kondziolka D, Lunsford LD, Coffey RJ, et al: Stereotactic radiosurgery of meningiomas. J Neurosurg 1991;74:552–559.
12 Kondziolka D, Lunsford LD: Radiosurgery of meningiomas. Neurosurg Clin North Am 1992;3:219–230.
13 Kondziolka D, Lunsford LD, Flickinger JC: Stereotactic radiosurgery for benign intracranial tumors. Concepts Neurosurg 1995;7:144–153.
14 Linskey ME, Martinez AJ, Kondziolka D, et al: The radiobiology of human acoustic schwannoma xenografts after stereotactic radiosurgery evaluated in the subrenal capsule of athymic mice. J Neurosurg 1993;78:645–653.
15 Linskey ME, Flickinger JC, Lunsford LD: Cranial nerve length predicts the risk of delayed facial and trigeminal neuropathies after acoustic tumor stereotactic radiosurgery. Int J Radiat Oncol Biol Phys 1993;25:227–233.
16 Lunsford LD, Witt T, Kondziolka D, et al: Stereotactic radiosurgery for anterior skull base tumors. Clin Neurosurg 1994;42:99–118.
17 Simpson D: The recurrence of intracranial menigiomas after surgical treatment. J Neurol Neurosurg J Psychiatry 1957;20:22–39.
18 Tishler RB, Loeffler JS, Lunsford LD, et al: Tolerance of the cranial nerves of the cavernous sinus to radiosurgery. Int J Radiat Oncol Biol Phys 1993;27:215–221.

Douglas Kondziolka, MD, University of Pittsburgh Medical Center, Suite B-400,
Department of Neurological Surgery, 200 Lothrop Street, Pittsburgh, PA 15213 (USA)
Tel. (412) 647 6782, Fax (412) 647 0989

Lunsford LD, Kondziolka D, Flickinger JC (eds): Gamma Knife Brain Surgery.
Prog Neurol Surg. Basel, Karger, 1998, vol 14, pp 114–127

....

Gamma Knife Radiosurgery for Pituitary Tumors

Thomas C. Witt [a], *Douglas Kondziolka* [b], *John C. Flickinger* [b],
L. Dade Lunsford [b]

[a] Section of Neurosurgery, Indiana University, Indianapolis, Ind.; and
[b] Department of Neurological Surgery, University of Pittsburgh, Pa., USA

Pituitary adenomas produce adverse symptoms by compression of adjacent cranial nerves, brain, and/or hypersecretion of hormones. Although prolactin-secreting tumors can often be successfully controlled medically by administration of dopamine agonists, microsurgical resection through either a transsphenoidal or transcranial route has been the primary method for managing patients with these tumors [1, 2]. With current microsurgical techniques, the risk of a serious complication is relatively low but not insignificant. A recent survey of neurosurgeons in the United States showed that transsphenoidal surgery for pituitary tumors is associated with a 3.9% risk of CSF leak, a 1.8% risk of new visual deficit, a 19.4% risk of pituitary insufficiency, and a 0.9% risk of death [3]. The risk of residual or recurrent tumor following microsurgical resection is not insignificant. Even when resection is performed by experienced surgeons, failure to achieve remission occurs in at least 15% of cases [4–6]. If a second transsphenoidal operation is performed, the success and complication rates are significantly less favorable. Laws et al. [7] reviewed the outcomes of 158 patients who underwent transsphenoidal surgery for lesions that had failed to respond to previous surgical, radiotherapeutic, or medical treatments. They found that the risks of serious complications as well as the risk of operative mortality were much higher in this group of patients. In addition, a successful resolution of endocrinopathy was achieved in only 56% of patients with acromegaly, 36% of patients with prolactinomas, and 25% of patients with ACTH-producing tumors. Nearly one half of the patients in that series still required some form of subsequent therapy. Other methods are needed for adjunctive treatment of patients with recurrent or residual disease following microsurgery and for primary treatment of patients unwilling or unable to tolerate the risks associated with microsurgery.

Fractionated radiation therapy has been the conventional method of treatment for unresectable pituitary adenomas. Rates of tumor growth control have been reported to vary from 76 to 97% [8–14]. Fractionated radiotherapy has been less successful (38–70%) in reducing hypersecretion of hormones by tumors [13–17]. A recent report by Estrada et al. [18], however, showed that 25 out of 30 patients (83%) with Cushing's disease had normalization of their pituitary-adrenal axis following 50 Gy of fractionated radiation at a mean follow-up of 42 months. Complications of fractionated radiation include a relatively high rate of hypopituitarism (12–100%) [9–12, 14, 15, 17, 19–21] and low but still significant risks of optic neuropathy (1–2%) [8, 10, 14, 17, 20] and induction of a secondary tumor (2.7% actuarial incidence at 15 years) [22].

The concept of stereotactic radiosurgery was introduced in 1951 by Professor Lars Leksell [23] as a noninvasive method for the precise destruction of intracranial targets using a single high dose of ionizing radiation. Many of the patients with pituitary adenomas who were treated early in the era of focused irradiation were treated with charged particles. In 1968, the first pituitary tumor patient was treated with the Gamma knife. In this report, we present our results using stereotactic radiosurgery with the Gamma Knife to manage patients with pituitary adenomas.

Techniques

Over a 10-year interval, 87 patients with pituitary adenomas were treated by stereotactic radiosurgery with the 201 source cobalt-60 Leksell Gamma knife (Elekta Instruments, Atlanta, Ga.). The age of the patients varied from 9 to 88 years (mean 46). Thirty-two of the patients (37%) were male, and 55 were female (63%). Twenty-four patients (28%) had endocrine-inactive tumors. There were 29 patients (33%) with ACTH-secreting tumors, 20 patients (23%) with growth hormone-secreting tumors, 12 patients (14%) with prolactinomas, and 2 patients (2%) with tumors producing both growth hormone and prolactin (table 1). Ten patients (11%) received stereotactic radiosurgery as the initial treatment modality for their tumors (fig. 1). Of the 77 patients who had been treated by microsurgery prior to radiosurgery, 74 (96%) had a transsphenoidal tumor resection and 17 (22%) had a transcranial approach to the tumor; 28 patients (36%) had more than one microsurgical procedure. Eighteen patients (23%) who had residual or recurrent tumor after microsurgery also had fractionated radiotherapy (mean dose 45.6 Gy) prior to radiosurgery. Four of the

Table 1. Types of pituitary tumors for radiosurgery

Tumor type	Patients	
	n	%
Nonsecreting adenoma	24	28
ACTH-secreting	29	33
GH-secreting	20	23
PRL-secreting	12	14
Mixed GH/PRL-secreting	2	2

patients with ACTH-producing adenomas also had bilateral adrenalectomy before radiosurgery.

First, patients had the Leksell Model G stereotactic frame applied using sterile technique, local anesthesia, and, when necessary, mild intravenous sedation. Children under the age of 14 had the procedure performed under general anesthesia. After frame application, a high-resolution stereotactic imaging study was obtained. Computed tomography was performed in 15 patients (17%), and magnetic resonance images were obtained in 72 (83%). On these studies, the tumor was seen to invade the cavernous sinus in 50 patients (57%). The mean tumor volume was 3.2 ml and varied from 0.22 to 14.1 ml.

Images were transferred into a computer for high-precision dose planning. Multiple isocenters of radiation using the 4-, 8-, 14- and 18-mm collimators were combined to create a therapeutic isodose line that matched the borders of the tumor (fig. 2) [24, 25]. The 50% isodose curve was used to enclose the tumor in the majority (86%) of cases since the slope of radiation fall-off is steepest at this point. Six tumors were irradiated at the 60% isodose, and the remaining tumors were treated at the 30% (n=2), 40% (n=2), 70% (n=1) and 80% (n=1) isodose curves. Maximum radiation dose and dose to the tumor margin were selected in order to minimize the risk of visual deficit and provide the highest potential for growth control and normalization of hormone production. These doses were chosen based on tumor volume and previous history of radiotherapy [26, 27]. The mean radiation dose to the tumor margin was 19.2 Gy (range 9.6–30). The mean maximum tumor dose was 37.9 Gy (range 20–60). When necessary, individual collimators within each helmet were plugged to shift peripheral isodose curves away from the optic nerve, chiasm, or tract in order to limit the dose to the optic apparatus to less than 9 Gy [28–30]. Customized beam-blocking patterns facilitated by GammaPlan software have made sellar and parasellar radiosurgery planning much easier.

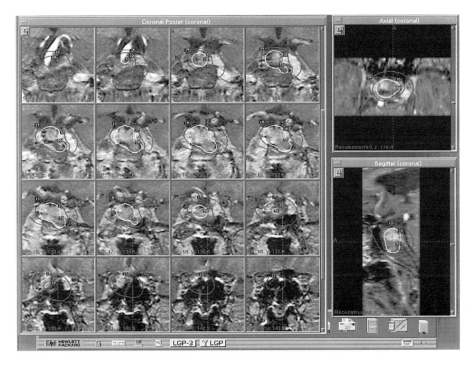

Fig. 1. MR-based Gamma knife radiosurgery dose plan in a 79-year-old woman with acromegaly. Radiosurgery was her first surgical procedure and was performed with one 14-mm and one 8-mm isocenter combined to deliver a maximum dose of 50 Gy. The 50% isodose line encircles the tumor on the coronal images. Axial and sagittal images are shown on the right. A beam-plugging pattern was used to restrict dose away from the optic chiasm.

Tumor Growth Control

Follow-up was available in 71 out of 87 patients (82%) over a period of 3–103 months (mean 32). Growth control of the tumor enclosed within the prescription isodose line was achieved in 67 patients (94%); tumor volume decreased in 33 patients (46%) and was unchanged in 34 patients (48%). Tumor volume increased in 4 patients (6%) despite radiosurgery. Three of these 4 patients had had previous microsurgery (once, twice, and four times) as well as fractionated radiotherapy. The fourth patient was an 88-year-old woman with hypertension and atrial fibrillation who presented with visual dysfunction from an endocrine-inactive macroadenoma. Because of her age and medical problems, she was considered to be a poor candidate for microsurgery. She had stereotactic radiosurgery, and the tumor received a maximum dose of 33.3 Gy with 10 Gy to the 30% isodose at the border. Three months following

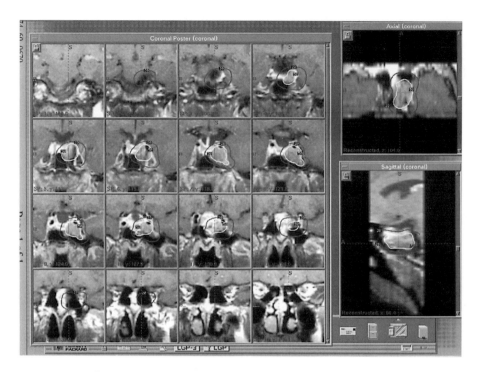

Fig. 2. Radiosurgery plan in a 58-year-old woman with acromegaly who had undergone prior transsphenoidal resection. The dose plan was created using one 14-mm, four 8-mm, and one 4-mm isocenter and a customized beam-blocking pattern. A maximum dose of 50 Gy was administered to this tumor in the left cavernous sinus.

radiosurgery, her vision continued to decline, and an MRI showed a very slight expansion of the volume of the tumor associated with decreased signal intensity in the central portion of the tumor. She had an endoscopic transsphenoidal removal of the tumor followed by significant improvement in vision postoperatively. Two patients with invasive prolactinomas developed new remote nodules of tumor in the cavernous sinus and along the optic tract and received second radiosurgical procedures. One of these tumors decreased in size and the other one was stable at 3 and 5 months, respectively, after the second Gamma knife procedure.

Cushing's Disease

In the group of patients with ACTH-producing tumors, endocrinologic follow-up was available in 21 out of 29 cases (72%). Normalization of pituitary-

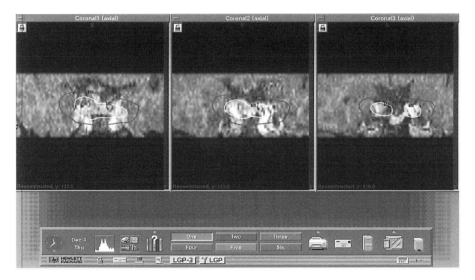

Fig. 3. Coronal MR images used for dose planning in a 35-year-old man with acromegaly show a plan created with seven 8-mm and one 4-mm isocenters as well as a customized blocking pattern to deliver a maximum dose of 40 Gy. The 20% isodose line remains below the optic chiasm and nerves.

adrenal function occurred in 11 of these 21 patients (52%) over a period of 4–39 months (mean 16). Normal function was defined by fasting a.m. cortisol (n = 6), 24-hour urinary free cortisol (n = 6), and/or serum ACTH levels (n = 4). Two of the patients with normal levels are also taking ketoconazole to limit hormone synthesis within the adrenal gland. Hormone levels in 3 out of 21 patients (14%) decreased by over 50% at a mean follow-up of 20 months (range 6–31) after radiosurgery; 1 of these patients is also still taking ketoconazole. Seven patients in this group (33%) did not have any significant reduction in ACTH or cortisol production. Six of these 7 patients were evaluated at a mean of 16 months (range 4–39) after radiosurgery. An ectopic ACTH-producing tumor was discovered in the posterior fossa of the seventh patient 72 months following radiosurgery.

Acromegaly

Endocrinologic follow-up was available in 16 out of 20 (80%) acromegalic patients treated with the Gamma knife (fig. 3). Fourteen of these 16 patients had both growth hormone and somatomedin-C levels tested; 2 patients only had analysis of somatomedin-C, and 1 patient was evaluated only by growth hormone levels. Growth hormone levels became normal in 10 out of 14 (72%) of patients at a mean follow-up of 29 months (range 14–52). One of these patients continues to take octreotide and 1 takes bromocriptine. Three out of 14 patients (21%) had a greater than 50% reduction in growth hormone levels by 37 months (range 13–72) following radiosurgery. The level of growth hormone in 1 patient in this group (7%) had not fallen by 15 months after radiosurgery. Reduction in somatomedin-C levels did not occur to the same degree as the reduction in growth hormone secretion. Somatomedin-C levels became normal in only 4 out of 15 patients (27%) at a mean follow-up of 31 months (range 12–52); 1 of these patients was also taking octreotide. An additional 2 out of 15 patients (13%) had a greater than 50% decrease in somatomedin-C levels by 58 months (range 44–72) after radiosurgery. In 9 of 15 patients (60%), somatomedin-C levels remained unchanged despite a mean postradiosurgery time interval of 32 months (range 15–88). Two of the patients who experienced a greater than 50% reduction in either growth hormone or somatomedin-C levels had a second radiosurgical operation in order to achieve that degree of hormone control.

Prolactinoma

Seven out of 12 patients (58%) with prolactinomas had follow-up endocrine testing (fig. 4). One patient who was taking bromocriptine had a normal prolactin level 4 months after radiosurgery. Prolactin levels dropped by 18–69% in 5 patients over 7–37 months (mean 17). One patient with a very aggressive tumor had an excellent response initially (65% decrease at 7 months) but then had progressive tumor growth, and by 29 months after radiosurgery, had a prolactin level greater than 10,000.

Two patients had mixed growth hormone-prolactin-secreting tumors. Prolactin and somatomedin-C levels were normal at 3 months following radiosurgery in 1 patient. In the other patient, somatomedin-C levels became normal and prolactin decreased by 40% 6 months after radiosurgery.

Morbidity of Radiosurgery

New or worsened neurological problems related to different factors developed in 9 out of the 71 patients (12.6%) who had adequate follow-up. In 4 of these patients (5.6%), new cranial neuropathies developed as a consequence of progressive tumor growth. As discussed above, 3 of these 4 patients had

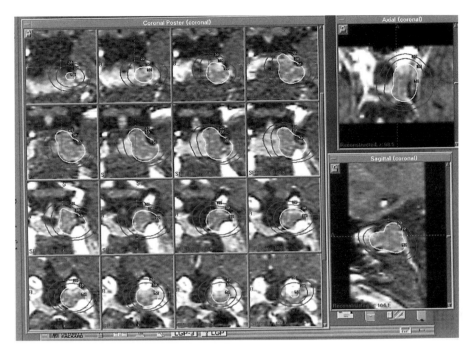

Fig. 4. Coronal MR images in a 35-year-old woman with a prolactinoma of the left cavernous sinus after prior transsphenoidal resection. Two 8-mm and one 4-mm isocenters were used to deliver a maximum dose of 50 Gy. No beam blocking was required using the model B Gamma knife.

tumors that had also progressively grown despite one or more microsurgical operations and despite fractionated radiotherapy. One patient (1.4%) with Cushing's disease died of central pontine myelinolysis after undergoing transsphenoidal microsurgery 13 months after radiosurgery because the tumor was still hypersecreting ACTH. Four patients (5.6%) developed new or increased neurologic deficits despite stable or decreased tumor size. These deficits are most likely related to effects of radiation. One patient who had previously had two microsurgical procedures and radiation therapy developed some enhancement of his hypothalamus 11 months after radiosurgery and, 5 months later, had a seizure and died. Another patient who had undergone both transcranial and transsphenoidal microsurgery plus radiation therapy had worsening of a pre-existing visual acuity deficit in one eye and of a visual field deficit in the other eye. Computed tomography was the imaging study performed in these latter two for stereotactic guidance. A third patient who had diabetes as well as a growth hormone-secreting adenoma invading the cavernous sinus reported a 25% decrease in sensation in the V-1 and V-2 divisions of his left trigeminal nerve; his corneal reflex remained intact. One patient with a prolactinoma involving the cavernous sinus who had decreased visual acuity and decreased peripheral vision in the left eye prior to radiosurgery suddenly became blind in the left eye 14 months after receiving 14 Gy to the 50% isodose line of her tumor and a maximum dose of 8 Gy to her optic nerve. She had symptoms of amaurosis in this eye before losing her vision. She has subsequently recovered the ability to detect hand motion and primary colors. She also developed a new partial oculomotor nerve palsy and decreased sensation on the left side of her face following radiosurgery.

To the best of our knowledge, no patient has developed a deficiency of pituitary hormone production following radiosurgery. Despite our requests for information from referring endocrinologists, however, only a few patients have undergone long-term detailed assessment of pituitary function other than measurements of the hypersecreted hormone.

Clinical Expectations after Radiosurgery

The results we have obtained using Gamma knife stereotactic radiosurgery for growth control of pituitary adenomas are similar to our results for control of other benign tumors such as meningiomas and vestibular schwannomas [31, 32]. This high rate of growth control has been particularly important for pituitary adenomas because microsurgery and radiation therapy had already failed to achieve this goal in a high percentage of these patients. Ganz et al. [33] also reported successful control of tumor growth in 14 out of 15 patients who were treated with the Gamma knife and followed for a minimum of 18 months. Microsurgery is still clearly the treatment of choice when visual function is compromised from mass effect by a macroadenoma. Our one attempt to prevent visual deterioration in an elderly patient with a macroadenoma was a failure. Other than this patient, the only failures of radiosurgery occurred in 3 patients with biologically aggressive tumors that had recurred and progressed despite fractionated radiation and one or multiple microsurgical procedures. Two patients who had had microsurgery prior to radiosurgery developed new nodules of tumor outside of the original tumor site and required a second Gamma knife operation. Although this type of occurrence was rare, it emphasizes the need for high-resolution neuroimaging studies utilizing optimal sequences including fat suppression techniques to distinguish tumor from packing material and scar tissue.

Although failure to achieve endocrinologic remission is not uncommon after microsurgery, an immediate reduction of hormone levels can be achieved in the majority of cases by tumor removal. Since there is a significant degree of morbidity and mortality associated with prolonged elevation of serum growth hormone or cortisol, this expedient resolution of endocrinopathy makes microsurgery the initial treatment of choice. In our experience as well as the experience of others, the success rates of radiosurgery in normalizing excessive hormone production can only be described as fair. Radiosurgery for endocrine-active pituitary tumors was first performed in the late 1950s with charged particles. At the Lawrence Berkeley Laboratory, Lawrence, Tobias and co-workers [34] primarily used the plateau range of helium ions to deliver doses of 50–150 Gy in four fractions to ACTH- and prolactin-producing tumors and 30–50 Gy to patients with acromegaly. In a cohort of 234 patients out of a group of 318 patients with acromegaly, mean growth hormone levels fell by 70% in the first year after treatment, became normal at a mean of 4 years, and remained normal for over 10 years. Mean basal cortisol levels became normal after 1 year in a cohort of 44 patients out of a total group of 83 patients with Cushing's disease. Seventeen patients with Nelson's syndrome experienced a marked fall in ACTH levels but rarely to normal values. Prolactin levels became normal in 12 out of 20 patients by 12 months after treatment. Approximately 1 out of 3 patients in the Berkeley series had some degree of anterior pituitary insufficiency. Additional complications in the 318 patients treated for acromegaly included 3 patients with seizures, 3 with extraocular muscle disorders, 2 with visual field deficits, and 2 with temporal lobe injuries.

At the Harvard Cyclotron Laboratory, Kjellberg and Kliman [35] used the Bragg peak of protons to treat approximately 600 patients with acromegaly and 175 patients with Cushing's disease. Remission, defined as serum growth hormone < 10 ng/ml, was achieved in 60% of patients at 24 months and 80% of patients at 48 months. In an early report of 22 patients with acromegaly, 3 patients had intermittent mild diplopia, 1 had anterior pituitary insufficiency, and 1 became blind in one eye and transiently confused. Normalization of clinical and laboratory cortisol function occurred in 65% of a group of 36 patients with Cushing's disease and another 20% had significant improvement.

Stereotactic radiosurgery as well as fractionated stereotactic radiotherapy has also been performed for pituitary adenomas with modified linear accelerators. In 26 patients who were followed for a median of 19 months after LINAC radiosurgery, no significant endocrine response was seen in patients with ACTH- or prolactin-producing tumors, and although growth hormone levels dropped in patients with acromegaly, these levels did not become normal [36]. The median dose delivered to the tumor margin in this series was only 14.5 Gy. The Joint Center for Radiation Therapy in Boston included 17 patients with pituitary adenomas in their initial reports on their use of fractionated stereotactic radiotherapy [37]. Although there were no complications and no tumor growth identified in this group, the follow-up period was short (3–16 months).

Other Gamma Knife Series

Results of Gamma knife radiosurgery for pituitary tumors have also been reported by Rahn et al. [38], Ganz et al. [33] and Park et al. [39]. Rahn et al. [38] achieved a remission rate of 82% in patients with Cushing's disease treated in the pre-MRI era and a remission rate of 100% in 8 patients who had an MRI study for target localization. Rahn et al. also reported a significant reduction of excessive hormone levels in 21 patients with acromegaly and 5 patients with prolactinomas. The incidence of pituitary insufficiency in the Stockholm series varied from 13 to 20%. All of the cases of pituitary insufficiency occurred when pneumoencephalography or CT was used for targeting and when up to four individual radiosurgical procedures were performed on the same patient for persistent endocrinopathy. New endocrinologic deficits were also more common in children. Using lower doses, Ganz et al. [33] reported successful growth control in 14 out of 15 patients and a decrease in excessive hormone levels in 93% of patients but an endocrinological cure in only 21%. Two of his patients required surgery for persistent endocrinopathy 17 and 24 months following radiosurgery despite the fact that the tumors had decreased in size. None of the patients in the series by Ganz et al. developed anterior pituitary insufficiency. Park et al. [39] performed 28 radiosurgical operations in 27 patients and reported normalization of growth hormone in 4 out of 7 patients, normalization of ACTH in 2 out of 5 patients, and a greater than 50% decrease in prolactin levels in 5 out of 7 patients with no neurologic or endocrine complications.

One of the most significant benefits of stereotactic radiosurgery is the extremely low incidence of pituitary insufficiency associated with this treatment modality. The relative resistance of normal pituitary tissue to single high doses of radiation was recognized early by Lawrence and co-workers [34] as well as Backlund et al. [41]. Patients with metastatic breast cancer were treated with doses of over 200 Gy to the center of the pituitary gland. When autopsies were performed on some of these patients within a few months of treatment, a sharp border between the necrotic area and a rim of normal tissue could be identified. Backlund et al. [4] determined that the necrosis began at a dose of 185 Gy.

With a mean clinical and radiographic follow-up of 32 months in our patients and 1 patient followed for over 8 years, we have not identified any patient who developed a new endocrine deficit following Gamma knife radiosurgery. Unfortunately, our supporting evidence for this statement mainly consists of the absence of reports of hormone deficiencies and the fact that patients have not needed to begin endocrine replacement therapy. The longest specific postradiosurgery documentation of normal pituitary function in our series has been 74 months. Hypopituitarism may develop between 6 months and 10 years following fractionated radiation therapy for pituitary tumors but most commonly occurs between 2 and 5 years. Therefore, despite our relatively limited follow-up in small numbers of patients compared to the decades of follow-up in hundreds of patients treated with fractionated radiation, the incidence of pituitary insufficiency following radiosurgery, even if it is not actually zero, appears to be significantly lower than the incidence following radiotherapy.

Stereotactic radiosurgery also is associated with a low rate of neurologic morbidity. Three patients (4%) developed worsening of pre-exisiting visual deficits unrelated to tumor growth. Two of these complications occurred in patients who were treated early in our series and whose treatment plans were based on CT images. The optic neuropathy that increased in the patient who had an MRI-based treatment plan may have had an indirect vascular rather than a direct axonal etiology. In 50 patients who had tumor involving the cavernous sinus, only 2 (4%) developed an alteration of facial sensation. One of these 2 patients also developed a partial third nerve palsy. These results support our experience that the cranial nerves traversing the cavernous sinus are highly tolerant of the doses used in tumor radiosurgery.

In summary: Microsurgery remains the treatment of choice for pituitary tumors when there is compression of the optic apparatus or when a rapid reduction in excessive hormone level is desired. For residual or recurrent tumors that are >2 or 3 mm away from optic structures, however, stereotactic radiosurgery provides growth control and long-term endocrine control superior to that of repeat microsurgery. Also, although the rates of growth control with radiosurgery are similar to the best results reported in radiotherapy series, the risk of subsequent hypopituitarism is significantly lower following radiosurgery and the risk of secondary tumor formation is not reported. Improvement of the results of radiosurgery in the future will depend on development of agents that sensitize neoplastic and endocrine-hyperactive tissue to radiation and other drugs that will protect adjacent neural structures.

References

1 Jaquet P: Medical therapy of prolactinomas. Acta Endocrinol 1993;129(suppl 1):31–33.
2 Wilson CB: The role of surgery in the management of pituitary tumors. Neurosurg Clin North Am 1990;1:139–159.
3 Ciric I, Ragin A, Baumgartner C, Pierce D: Complications of transsphenoidal surgery: Results of a national survey, review of the literature, and personal experience. Neurosurgery 1997;40: 225–237.
4 Ciric I, Mikhail M, Stafford T, Lawson L, Garces R: Transsphenoidal microsurgery of pituitary macroadenomas with long-term follow-up results. J Neurosurgy 1983;59:395–401.
5 Tindall GT, Herring CJ, Clark RV, Adams DA, Watts NB: Cushing's disease: Results of trans-sphenoidal microsurgery with emphasis on surgical failures. J Neurosurg 1990;72:363–369.
6 Tindall GT, Oyesiku NM, Watts NB, Clark RV, Christy JH, Adams DA: Transsphenoidal adenomec-tomy for growth hormone-secreting pituitary adenomas in acromegaly: Outcome analysis and determinants of failure. J Neurosurg 1993;78:205–215.
7 Laws ER Jr, Fode NC, Redmond MJ: Transsphenoidal surgery following unsuccessful prior therapy. J Neurosurg 1985;63:823–829.
8 Flickinger JC, Nelson PB, Martinez AJ, Deutsch M, Taylor F: Radiotherapy of nonfunctional adenomas of the pituitary gland. Cancer 1989;63:2409–2414.
9 Salinger DJ, Brady LW, Miyamoto CT: Radiation therapy in the treatment of pituitary adenomas. Am J Clin Oncol 1992;15:467–473.
10 McCollough WM, Marcus RB Jr, Rhoton AL, Ballinger WE, Million RR: Long-term follow-up of radiotherapy for pituitary adenoma: The absence of late recurrence after >4,500 Gy. Int J Radiat Oncol Biol Phys 1991;21:607–614.
11 Rush S, Cooper PR: Symptom resolution, tumor control, and side effects following postoperative radiotherapy for pituitary macroadenomas. Int J Radiat Oncol Biol Phys 1997;37:1031–1034.
12 Tsang RW, Brierley JD, Panzarella T, Gospodarowicz MK, Sutcliffe SB, Simpson WJ: Radiation therapy for pituitary adenoma: Treatment outcome and prognostic factors. Int J Radiat Oncol Biol Phys 1994;30:557–565.
13 Tsang RW, Brierley JD, Panzarella T, Gospodarowicz MK, Sutcliffe SB, Simpson WJ: Role of radiation therapy in clinical hormonally-active pituitary adenomas. Radiother Oncol 1996;41:45–53.
14 Zierhut D, Flentje M, Adolph J, Erdmann J, Raue F, Wannenmacher M: External radiotherapy of pituitary adenomas. Int J Radiat Oncol Biol Phys 1995;33:307–314.
15 Goffman TE, Dewan R, Arakaki R, Gordon P, Oldfield EH, Glutstein E: Persistent or recurrent acromegaly: Long-term endocrinologic efficacy and neurologic safety of post-surgical radiation therapy. Cancer 1992;69:271–275.
16 Howlett TA, Plowman PN, Wass JAH, Rees LH, Jones AE, Besser GM: Megavoltage pituitary irradiation in the management of Cushing's disease and Nelson's syndrome: Long-term follow-up. Clin Endocrinol 1989;31:309–323.
17 Rush SC, Newall J: Pituitary adenomas: The efficacy of radiotherapy as the sole treatment. Int J Radiat Oncol Biol Phys 1989;17:165–169.
18 Estrada J, Boronat M, Mielgo M, Magallon R, Millan I, Diez S, Lucas T, Barcolo B: The long-term outcome of pituitary irradiation after unsuccessful transphenoidal surgery in Cushing's disease. N Engl J Med 1997;336:172–177.
19 Littley MD, Shalet SM, Beardwell CG, Ahmed SR, Applegate G, Sutton ML: Hypopituitarism following external radiotherapy for pituitary tumors in adults. Am J Med 1989;79:145–160.
20 Fisher BJ, Gaspar LE, Noone B: Radiation therapy of pituitary adenoma: Delayed sequelae. Radiology 1993;187:843–846.
21 Tominaga A, Uozumi T, Arita K, Kurisu K, Yano T, Hirohata T: Anterior pituitary function in patients with nonfunctioning pituitary adenoma: Results of longitudinal follow-up. Endocr J 1995; 42:421–427.
22 Tsang RW, Laperriere NJ, Simpson WJ, Brierley J, Panzarella T, Smyth HS: Glioma arising after radiation therapy for pituitary adenoma. Cancer 1993;72:2227–2233.
23 Leksell L: The stereotaxic method and radiosurgery of the brain. Acta Chir Scand 1951;102:316–319.

24 Flickinger JC, Lunsford LD, Wu A, Maitz AH, Kalend AM: Treatment planning for gamma knife radiosurgery with multiple isocenters. Int J Radiat Oncol Biol Phys 1989;18:1495–1501.

25 Wu A, Lindner G, Maitz AH, Kalend AM, Lunsford LD, Flickinger JC, Bloomer WD: Physics of gamma knife approach on convergent beams in stereotactic radiosurgery. Int J Radiat Oncol Biol Phys 1990;18:941–949.

26 Flickinger JC, Lunsford LD, Kondziolka D: Dose prescription and dose-volume effects in radiosurgery. Neurosurg Clin North Am 1992;3:51–59.

27 Flickinger JC, Deutsch M, Lunsford LD: Repeat megavoltage irradiation of pituitary and suprasellar tumors. Int J Radiat Oncol Biol Phys 1989;17:171–175.

28 Flickinger JC, Maitz AH, Kalend A, Lunsford LD, Wu A: Treatment volume shaping with selective beam blocking using the Leksell gamma unit. Int J Radiat Oncol Biol Phys 1990;19:783–789.

29 Stephanian E, Lunsford LD, Coffey RJ, Bissonette DJ, Flickinger JC: Gamma Knife surgery for sellar and suprasellar tumors. Neurosurg Clin North Am 1992;3:207–218.

30 Duma CM, Lunsford LD, Kondziolka D, Harsh GR, Flickinger JC: Stereotactic radiosurgery of cavernous sinus meningiomas as an addition or alternative to microsurgery. Neurosurgery 1993;32: 699–705.

31 Kondziolka D, Lunsford LD, Flickinger JC: Stereotactic radiosurgery of meningiomas; in Lunsford D, Kondziolka D, Flickinger JC (eds): Gamma Knife Brain Surgery. Prog Neurol Surg. Basel, Karger, 1998, vol 14, pp 104–113.

32 Lunsford LD, Kondziolka D, Pollock BE, Flickinger JC: Gamma knife radiosurgery for acoustic neuromas; in Lunsford LD, Kondziolka D, Flickinger JC (eds): Gamma Knife Brain Surgery. Prog Neurol Surg. Basel, Karger, 1998, vol 14, pp 89–103.

33 Ganz JC, Backlund EO, Thorsen FA: The effects of gamma knife surgery of pituitary adenomas on tumor growth and endocrinopathies. Stereotact Funct Neurosurg 1993;61(suppl 1):30–37.

34 Levy RP, Fabrikant JI, Frankel KA, Phillips MH, Lyman JT, Lawrence JH, Tobias CA: Heavy-charged-particle radiosurgery of pituitary gland: Clinical results of 840 patients. Stereotact Funct Neurosurg 1991;57:22–35.

35 Kjellberg RN, Kliman B: Lifetime effectiveness – A system of therapy for pituitary adenomas, emphasizing Bragg peak proton hypophysectomy; in Linfoot JA (ed): Recent Advances in the Diagnosis and Treatment of Pituitary Tumors. New York, Raven Press, 1979, pp 269–288.

36 Voges J, Sturm V, Deuss U, Traud C, Treuer H, Schlegel W, Winkelmann W, Muller RP: LINAC radiosurgery in pituitary adenomas: Preliminary results. Acta Neurochir 1996;65(suppl):41–43.

37 Dunbar SF, Tarbell NJ, Kooy HM, Alexander E III, Black PM, Barnes PD, Goumnerova L, Scott RM, Pomeroy SL, La Vally B, Sallan SE, Loeffler JS: Stereotactic radiotherapy for pediatric and adult brain tumors: Preliminary report. Int J Radiat Oncol Biol Phys 1994;30:531–539.

38 Rahn T, Thoren M, Werner S: Stereotactic radiosurgery in pituitary adenomas; in Faglia G, Beck-Peccoz P, Ambrosi B, Travaglini P, Spada A (eds): Pituitary Adenomas: New Trends in Basic and Clinical Research. New York, Excerpta Medica, 1991, pp 303–312.

39 Park YG, Chang JW, Kim EY, Chung SS: Gamma Knife surgery in pituitary microadenomas. Yonsei Med 1996;37:165–173.

40 Ganz JC, Aanderud S, Mork SJ, Smievoll AI: Tumour volume reduction following gamma knife radiosurgery: The relationship between x-ray and histological findings. Acta Neurochir 1994; 62(suppl):39–42.

41 Backlund EO, Rahn T, Sarby B, de Schryver A, Wennerstrand J: Closed stereotaxic hypophysectomy by means of 60-cobalt gamma radiation. Acta Radiol Ther Phys Biol 1972;11:545–555.

Thomas C. Witt, MD, Section of Neurosurgery, Indiana University,
Indianapolis, IN 46202 (USA)
Tel. (317) 274 8118, Fax (317) 278 3185

Lunsford LD, Kondziolka D, Flickinger JC (eds): Gamma Knife Brain Surgery.
Prog Neurol Surg. Basel, Karger, 1998, vol 14, pp 128–144

....

Stereotactic Radiosurgery for Other Skull Base Lesions

N. Muthukumar [a], *D. Kondziolka* [b], *L. D. Lunsford* [b]

[a] Department of Neurosurgery, Madurai Medical College, Madurai, India; and
[b] Department of Neurological Surgery, University of Pittsburgh, Pittsburgh, Pa., USA

The skull base presents a substantial challenge to surgical access. Problems are posed by the critical neurovascular structures that lesions in the region involve, by the depth of the surgical field required, and the obstruction of operative exposure by important soft tissue structures of the head and neck and by bone that is structurally important, dense, and irregular and that itself invests crucial neural and vascular entitities [1]. The complications of skull base surgery include: neural and vascular injury, CSF leakage, infection and cosmetic deformities. Given these problems, surgical treatment of lesions of the skull base should be undertaken only with unequivocal necessity and extensive forethought and preparation.

Skull base surgery has evolved through certain phases of development: early skepticism; fanaticism with overapplication; reconciliation, with reports of morbidity and recurrence; and maturation, in which reasonable goals and expectations translate into reasonable application [2]. However, modern skull base surgery has made hitherto inaccessible lesions both accessible and resectable. As we move into the second decade of modern skull base surgery, the emphasis has shifted from designing newer approaches to one that aims at preserving neural function.

Neurosurgical progress has been described as a 'thorny road to minimally invasive techniques' [3]. Major neurosurgical advances in the past two decades have consistently aimed at minimizing the invasiveness of neurosurgical intervention while enhancing the safety and effectiveness of surgical objectives [4]. A less invasive approach that achieves the same goals and objectives is necessarily of better quality than a more invasive one. Stereotactic radiosurgery using the Leksell Gamma knife represents one of the minimally invasive techniques that is emerging as an adjunct or alternative to conventional surgical procedures in the management of many skull base lesions.

Since the installation of the first Leksell Gamma knife unit in North America in Pittsburgh in 1987, we have had the opportunity to treat many skull base lesions with success. In this chapter we report our results with jugular foramen schwannomas, chordomas and chondrosarcomas and trigeminal schwannomas. The actual technique of Gamma knife radiosurgery has been dealt with in the earlier chapters.

Jugular Foramen Schwannomas

Schwannomas arising from the glossopharyngeal, vagus or accessory cranial nerves are rare, constituting only 2.9–4% of all intracranial schwannomas [5, 6]. Despite recent advances in skull base surgical techniques, the surgical challenge posed by jugular foramen region schwannomas (JFS) remains formidable and associated with significant cranial nerve morbidity [6–22]. In many instances, complete removal is not possible, recurrent growth is frequent [6], and reoperation for recurrent tumors is likely to increase the cranial nerve deficits [5].

In this report, we review 16 patients who had stereotactic radiosurgery for primary or adjuvant management of jugular foramen schwannomas. This is the largest surgical series reported to date.

Clinical Materials and Methods
Between May 1990 and December 1996, 16 patients (11 men and 5 women) with JFS underwent stereotactic radiosurgery with the 201-source cobalt-60 Gamma knife at the Center for Image-Guided Neurosurgery, University of Pittsburgh. All but 3 patients had undergone between 1 and 6 microsurgical resections (average 1.3 operations/patient). In the remaining three patients primary radiosurgical treatment was chosen because of either associated medical conditions, advanced age, and/or the goal of preservation of intact lower cranial nerve function. No patient had received prior fractionated external beam radiation therapy. The duration of symptoms prior to radiosurgery varied from 1 to 84 months (median 23). Patients were accepted for radiosurgery if the average tumor diameter was <35 mm and if primary or additional microsurgical resection was rejected either by the physician, patient, or both. The presenting neurological deficits of the patients are listed in table 1. The eighth cranial nerve was the most frequently involved followed by the ninth and tenth cranial nerves. Pellet et al.'s [21] modification of Kaye et al.'s [16] classification was used to grade these tumors (table 2). Thirteen patients (81.5%) had type A tumors (predominantly intracranial tumors with minimal extension into the jugular foramen), 1 patient (6.25%) had type C tumor (predominantly extracranial with extension into the jugular foramen), and 2 patients (12.5%) had type D tumors (dumbbell-shaped tumor with both intra- and extracranial extensions).

Table 1. Cranial nerve deficits in patients with JFS

Patient No.	V	VI	VII	VIII	IX	X	XI	XII
1	–	–	+	+	–	–	–	–
2	–	+	+	–	–	+	–	+
3	–	–	+	+	+	+	–	+
4	–	–	+	+	+	+	–	–
5	–	–	–	+	–	–	–	–
6	–	–	–	+	–	–	–	–
7	–	–	–	+	–	–	–	–
8	–	–	–	–	+	+	+	+
9	–	–	–	+	–	–	–	–
10	–	–	–	–	–	–	–	–
11	–	–	+	–	+	+	+	+
12	+	–	+	+	+	+	+	+
13	–	–	–	–	+	+	–	+
14	–	–	–	+	–	–	–	–
15	–	–	–	+	+	+	–	–
16	+	–	–	–	+	+	–	–

+ = Present; – = absent.

Table 2. Neuroimaging findings in JFS

Type	Description	Patients, n
Type A	Primary intracranial tumor with minimal enlargement of the jugular foramen	13
Type B	Tumor primarily at the jugular foramen with intracranial extension	0
Type C	Primary extracranial tumor with extension into the jugular foramen	1
Type D	Dumbbell-shaped tumor with both intra- and extracranial components	2

Table 3. Dosimetry in JFS

Dosimetric values	Tumors, n
Tumor margin dose	
12 Gy	2
14 Gy	1
15 Gy	6
16 Gy	4
18 Gy	3
Isodose at margin	
50%	15
70%	1

Results

The maximum tumor dose ranged from 21 to 36 Gy (median 30). The minimum tumor dose (tumor margin dose) ranged from 12 to 18 Gy (median 15) (table 3). The 4-, 8-, 14-, and 18-mm collimators were used either separately or in combination. Multiple irradiation isocenters were required to treat 88% of the JFS because of their irregular and complex shapes [23].

The median clinical follow-up was 33 months (range 6–68). Five patients (31%) had clinical improvement during follow-up. All but 1 of the remaining patients remained stable. One patient had progression of the preradiosurgery symptoms 6 months after irradiation and underwent microsurgical resection. During follow-up, 7 patients (43.7%) showed a reduction in tumor size (fig. 1), 8 patients (50%) showed no evidence of tumor progression and 1 patient showed progressive increase in size of the tumor and underwent microsurgical resection (table 4). Six of the tumors which did not show tumor progression showed evidence of central necrosis (fig. 2).

None of the patients developed complications directly attributable to radiosurgery. One patient had progressive increase in tumor size and subsequently underwent resection. Among all other patients, none had worsening of cranial nerve or other neurological deficits.

Discussion

In view of the difficulties associated with surgical resection, and because of the good results after acoustic tumor radiosurgery [24–29], we began to treat JFS with Gamma knife radiosurgery. All except 3 of these patients were referred for radiosurgery after microsurgical resections. Many of these patients had developed additional lower cranial nerve deficits following microsurgical resection.

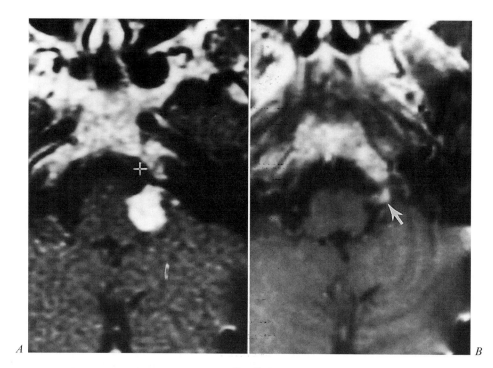

Fig. 1. Axial MR scans before (*A*) and following radiosurgery (*B*) for a JFS showing almost complete tumor regression (arrow).

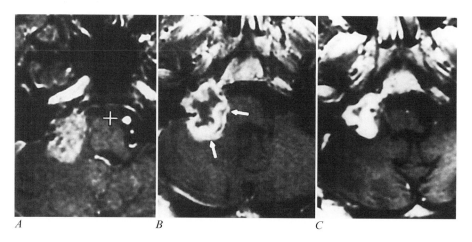

Fig. 2. Serial MR scans at radiosurgery (*A*), 6 months (*B*) and 18 months (*C*) later. Significant loss of central contrast enhancement was noted at 6 months with some expansion of the tumor diameters (arrows). Subsequently, tumor regression followed.

Neuroimaging change	Patients, n (%)
Tumor size reduced	7 (43.5%)
Tumor size unchanged	8 (50%)
Tumor size increased	1 (6.25%)

Table 4. Postradiosurgery imaging changes

Schwannomas have a number of characteristics that render them potentially well suited for radiosurgery: (1) they are well encapsulated; (2) they do not invade the brain; (3) they are readily defined by contrast-enhanced MR imaging, and (4) the steep radiation falloff achieved by radiosurgery can be conformed to the irregular tumor margins.

There have been few reports in the literature on radiosurgery for JFS [30, 31]. Pollock et al. [31] reported our initial 5 patients with JFS (these patients have been included in the present study with longer follow-up). Four of these patients showed absence of tumor growth during the follow-up period. No complications were noted specific to the radiosurgery. Since all these patients had prior microsurgical resections and cranial nerve deficits before radiosurgery, it was not possible to assess the response of the lower cranial nerves to this procedure. Kida et al. [30] reported their experience with radiosurgery for jugular foramen lesions, four of which were schwannomas. Three of the four tumors showed reduction in size during the mean follow-up period of 19 months. The remaining patient showed central tumor necrosis. No patient had evidence of radiation-induced edema surrounding the treatment site.

To date, no large radiosurgery series of JFS has been reported. In our experience, 7 (44%) patients showed imaging evidence of decreased tumor size, 8 (50%) showed no further tumor growth, and 1 showed evidence of tumor progression. Six of 8 patients with no tumor progression showed imaging evidence of central necrosis. Our prior experience with acoustic schwannomas has shown that tumors showing central necrosis often undergo delayed volume reduction [24].

Unlike microsurgical resection, radiosurgery does not remove a JFS; instead, the goal is long-term prevention of tumor growth accompanied by low surgical risk. The twin goals of radiosurgery are: (1) tumor control and (2) cranial nerve preservation. In all 3 patients for whom radiosurgery was carried out as the primary treatment modality, this goal was accomplished. The remaining patients presented with multiple cranial nerve palsied following microsurgical resections and hence it is difficult to assess the risk for new cranial nerve morbidity. However, in all except 1 patient, the first goal of tumor growth prevention was achieved. From our experience, we believe that tumor growth control is a satisfactory outcome, especially when associated with a low surgical risk of new or additional cranial nerve dysfunction. In view of the tendency of these tumors to recur after incomplete removal [6] and the difficulties encountered in preserving cranial nerve function in recurrent tumors [5], we believe that residual or recurrent JFS should be subjected to radiosurgery. Large tumors in patients with intact lower cranial nerve function should be debulked with the goal of preservation of cranial nerve function and the residual tumor treated with radiosurgery. Such a cooperative effort may improve the ultimate functional outcome of patients with JFS. In small and medium-sized schwannomas in elderly patients, patients who are medically unfit for microsurgical resection, those with intact lower cranial nerve function, and patients who refuse microsurgical resection, radiosurgery should be considered as an alternative to resection.

Our results indicate that Gamma knife radiosurgery is a safe primary or adjuvant treatment method for selected patients with JFS. We anticipate that the long-term rate of growth control will approximate that found with acoustic tumors.

Chordomas and Chondrosarcomas

Chordomas are rare neoplasms that represent approximately 2–4% of primary bone neoplasms [31, 32] and 0.1–0.2% of primary intracranial neoplasms [34, 35]. One-third of chordomas arise from the base of the skull [36, 37] and are presumed to arise from notochordal remnants [36, 38, 39]. Chondrosarcomas are also uncommon tumors that often arise at the base of the skull and are believed to arise from primitive mesenchymal cells or from embryonal rests of cartilaginous matrix of the cranium [36, 40]. Both these tumors have variable growth rates, infiltrate adjacent tissues, and often recur locally. Because total surgical resection is often not feasible, other management strategies must often be considered [33, 41, 42, 43].

Clinical Materials and Methods
Between August 1987 and April 1997, 15 patients (9 with chordomas and 6 with chondrosarcomas) underwent stereotactic radiosurgery with the Leksell Gamma knife at the Center for Image-Guided Neurosurgery, Pittsburgh. Patient age varied from 7 to 70 years (mean 38). There were 10 males and 5 females. The most common presenting symptom was diplopia (73%) and abducens nerve deficit was the most common presenting sign (86%). Other cranial nerve deficits encountered in these patients are listed in table 5. Eleven of these patients had undergone between 1 and 4 prior surgeries (mean 1.8 operations/patient). Two other patients had a trans-sphenoidal biopsy for confirmation of tumor

Table 5. Cranial nerve deficits in patients with chordomas and chondrosarcomas of the cranial base

Patient No.	III	IV	V	VI	VII	VIII	IX	X	XI	XII
1	+	+	+	+	−	−	+	−	−	+
2	+	−	−	+	−	−	−	−	−	−
3	+	−	−	+	−	−	−	−	−	−
4	−	−	−	+	−	−	−	−	−	−
5	−	−	−	−	−	−	−	−	−	−
6	−	−	+	+	+	+	−	−	−	−
7	−	−	+	+	+	+	+	+	+	−
8	−	−	−	+	−	−	−	−	−	−
9	−	−	−	+	−	−	−	−	−	−
10	+	+	+	+	−	−	−	−	−	−
11	+	−	−	+	−	+	−	−	−	−
12	−	−	−	+	−	−	−	−	−	−
13	−	−	−	+	−	−	−	−	−	−
14	−	+	−	+	−	−	−	−	−	−
15	−	−	−	−	−	−	−	−	−	−

+ = Present; − = absent.

histology. The remaining two patients underwent radiosurgery on imaging criteria alone. In these 2 patients, radiosurgery was selected for management of small tumors that had caused an isolated abducens deficit. Three of the 15 patients had undergone prior fractionated external beam radiotherapy prior to referral for treatment of the recurrent or uncontrolled local disease.

Results

The maximum tumor dose varied from 24 to 80 Gy (median 36). The minimum tumor dose (tumor margin dose) varied from 11.7 to 40 Gy (median 18). The 4-, 8-, 14- and 18-mm collimators were used either separately or in combination. The number of isocenters varied from 1 to 10 (average 4). Multiple isocenters were required to treat 73% of the cranial base chordomas and chondrosarcomas because of their irregular and complex shapes.

The median clinical follow-up was 40 months (range 6–84). During the period of follow-up, 8 (53.3%) improved clinically, 3 (20%) remained stable, and 4 (26.6%) died. One of the patients who remained stable clinically for 2 years showed imaging evidence of growth of the extracranial, nonirradiated portion of the tumor and subsequently underwent resection. Eight of the patients showed improvement of their cranial nerve deficits. There was no

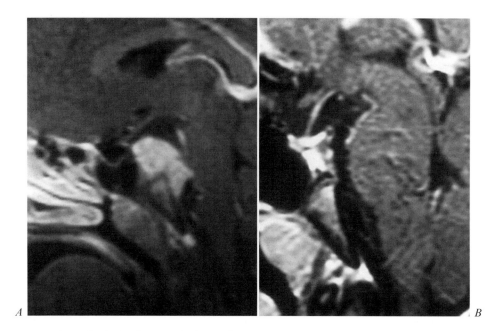

Fig. 3. Sagittal MR scans before radiosurgery (*A*) and 1 year later (*B*) in a 9-year-old boy with a clivus chordoma confirmed on needle biopsy. Complete tumor regression was identified and his abducens deficit resolved within 1 month.

morbidity directly attributable to radiosurgery. No patient had additional cranial or vascular or endocrine deficits following radiosurgery.

Five (33.3%) of patients showed imaging evidence of decrease in tumor size (fig. 3), 5 (33.3%) showed no further tumor, 4 (26.6%) showed increase in tumor size and 1 (6.6%) showed growth of the extracranial, nonirradiated portion of the tumor. The tumors which did not show further growth showed evidence of patchy areas of decreased contrast enhancement (fig. 4). No patient had evidence of signal changes in the brainstem or imaging changes attributable to narrowing of major intracranial vessels.

Discussion

Radiosurgery is a potent radiobiologic approach for the treatment of these lesions. Since radiosurgery is performed in a single fraction, it offers the increased radiobiologic effect inherent in this type of radiation delivery. Proton or charged particles, combine a higher radiobiologic effect than photons with the safety of fractionation. However, the targeting is not stereotactic. Brachytherapy offers a longer irradiation time than other forms of radiation but does not allow truly conformal irradiation and requires an invasive procedure for

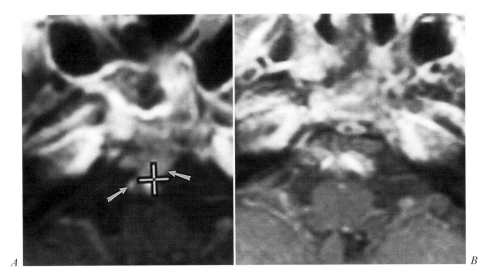

Fig. 4. Axial MR scans showing a clivus chordoma at the level of the medulla. At radiosurgery (*A*) the tumor is well defined (arrows). Two years later the mass showed decreased contrast enhancement and no growth (*B*).

seed placement. Although no comparative studies have been performed, we believe that stereotactic radiosurgery offers specific advantages for smaller chordomas and chondrosarcomas.

The aim of radiosurgery is to provide control of tumor growth while maintaining the integrity of the normal surrounding brain and cranial nerves. In these patients, radiosurgery was used to: (1) attempt to provide tumoricidal treatment in a single treatment session; (2) to spare critical brain structures, and (3) to reduce the morbidity reported after high-dose fractionated irradiation and proton beam therapy. In order to deliver high doses safely, tumors < 30 mm in maximum diameter and at least 5 mm from the optic chiasm were chosen. Multiple isocenters of irradiation or selective beam-blocking patterns were used to produce conformal isodose configurations that limit the dose to adjacent critical neural structures.

Overall, 73% of patients either improved clinically or remained stable following treatment and two-thirds either showed a reduction or stabilization of the tumor size on follow-up imaging. Radiosurgery for chordomas and chondrosarcomas of the cranial base is associated with no significant morbidity in our experience.

The failures in this series may be due to several reasons: (1) inherent biologically aggressive nature of the tumor treated; (2) failure to adequately estimate the overall tumor volume because of the limitations of imaging (early in the series when CT was used), and (3) inadequate dosage.

Since total removal is difficult to achieve in chordomas and chondrosarcomas of the cranial base [42, 43], residual tumor following surgery is common. These patients can be observed for regrowth of residual tumor. However, successive reoperations become progressively more difficult because the normal anatomy and the surgical planes are obliterated by fibrosis and the mass of the recurrent tumor. In this scenario, good local control can be achieved with stereotactic radiosurgery.

From our experience, we believe that small chordomas and chondrosarcomas of the cranial base in neurologically well-preserved patients can be safely and effectively controlled with stereotactic radiosurgery. In larger tumors, surgical debulking followed by radiosurgical treatment of the residual tumor seems to be a rational choice. Stereotactic radiosurgery can also help to provide a 'boost' to the tumor following conventional fractionated radiation therapy.

Trigeminal Schwannomas

Trigeminal schwannomas are uncommon lesions accounting for 0.07–0.36% of all intracranial tumors and 0.8–8% of all intracranial schwannomas [44–52]. The anatomical relationship of the trigeminal nerve along its intracranial course with the elements of the cerebellopontine angle, the petrous apex, the cavernous sinus, and the cranial nerves makes it very difficult to remove these tumors completely. In the era before microsurgery, total removal of trigeminal schwannomas was achieved in only 50% of cases [53] and was often associated with unacceptable mortality and morbidity [54]. With the advent of modern skull base approaches, the mortality and morbidity has been reduced considerably [49–53, 55]. However, even with the advent of modern microsurgical methods, complete removal may be difficult to achieve. In patients with residual tumors, patients who are medically unfit or unwilling to undergo major microsurgical procedures, alternative therapeutic strategies should be available. Gamma knife radiosurgery is an attractive adjunctive or alternative therapeutic option in such situations.

Clinical Materials and Methods
Between February 1989 and June 1996, 12 patients (8 men and 4 women) with trigeminal schwannomas underwent Gamma knife radiosurgery at the University of Pittsburgh. Six of the patients had undergone between 1 and 3 microsurgical resections (average 1.3 resections/patient) prior to radiosurgery. The remaining patients underwent radiosurgery on imaging criteria alone. These patients had primary radiosurgical treatment either because of the

Table 6. Neurological deficits in 12 trigeminal schwannomas

Patient No.	III	IV	V	VI	VII	VIII	IX	X	XI	XII	Other signs
1	–	–	+	–	–	–	–	–	–	–	–
2	–	–	+	–	–	–	–	+	–	–	–
3	–	–	+	+	–	–	–	–	–	–	–
4	+	–	+	+	–	–	–	–	–	–	–
5	–	–	+	–	–	–	–	–	–	–	–
6	–	–	–	–	–	–	–	–	–	–	–
7	–	–	+	–	–	–	–	–	–	–	–
8	+	–	+	–	–	–	–	–	–	–	–
9	–	–	+	–	–	–	–	–	–	–	–
10	–	–	+	+	+	+	–	–	–	–	–
11	–	–	+	–	–	–	–	–	–		
12	–	–	+	+	–	–	–	–	–	–	–

Roman numerals indicate corresponding cranial nerve.
+ = Present; – = absent.

associated medical conditions, advanced age, or refusal of the patient to undergo microsurgical resection. No patient had prior fractionated external beam radiotherapy.

The median age was 57 years (range 26–82). The presenting neurological deficits of the patients are listed in table 6. Symptoms and signs referrable to the trigeminal nerve were the most common.

Results

The maximum tumor dose varied from 24 to 40 Gy (mean 32.4). The tumor margin dose varied from 10.8 to 20 Gy (mean 16.2). Multiple irradiation isocenters were required to treat all but one lesion because of their irregular and complex shapes. The number of isocenters varied from 1 to 13 (average 5.17/patient). The tumor margin received a 50% isodose in 10 patients, 45% in 1 and 55% in 1 patient.

The median clinical follow-up was 2 years (range 8–49 months). Five patients showed improvement of clinical symptoms, 6 showed no change and 1 patient continued to have worsening of the symptoms. Two patients aged 82 and 75 years died 2 years and 9 months following radiosurgery due to unrelated cardiac illnesses. Both the patients had improvement of their symptoms and showed imaging evidence of tumor growth control during the last follow-up.

Five patients showed reduction in tumor size during the follow-up period (fig. 5), 6 showed no further tumor growth and 1 patient showed continued

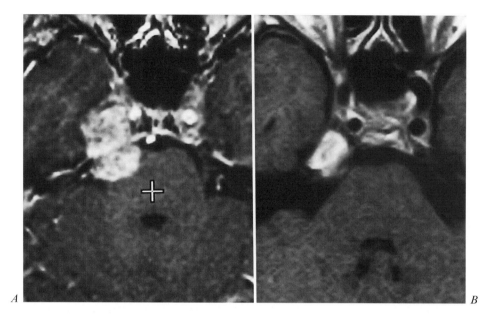

A   B

Fig. 5. Axial MR scans at radiosurgery (*A*) and 3.5 years later (*B*) showing significant regression of a right trigeminal schwannoma. Regression occurred from both the posterior fossa and cavernous sinus components.

tumor growth. Three of the 6 patients with tumor growth control showed imaging evidence of central necrosis.

Discussion

Despite their rarity, these lesions have attracted considerable attention in the literature mainly because of the difficulty in totally removing these tumors which is due to their location and their anatomic relationship to the surrounding vital neural and vascular structures [53]. Schisano and Olivecrona [54] reviewed the world literature and reported a mortality rate of 41% following treatment of these lesions. Arseni et al. [44] reported a mortality rate of 25% in cases published up to 1970. Prior to the advent of microsurgical era, total removal of these lesions could be achieved only in 50% of the cases [53]. However, with the advent of modern microsurgical and skull base techniques, radical removal of these lesions has become possible [49–51, 53, 55]. Nevertheless, the strategic location of these tumors and their adherence to vital vascular and neural structures continue to pose a threat of significant operative morbidity and frequently proven complete extirpation [49].

There have been few reports in the literature on radiosurgery for trigeminal schwannomas [31, 56]. Pollack et al. [31] reported our initial 6 patients with trigeminal schwannomas. Tumor growth control was achieved in all the patients. Three patients showed a reduction in tumor size and 3 showed no further growth of the tumor. There were no complications attributable to radiosurgery. Kida et al. [56] reported the results of 10 nonacoustic schwannomas, 5 of which were trigeminal schwannomas. Even though detailed breakup was not available for trigeminal schwannomas, their series had a 100% tumor growth control rate with no complications. Yamasaki et al. [57] reported a case of giant trigeminal schwannoma extending from the cavernous sinus to the cerebellopontine angle. The tumor was removed radically except for the portion within the cavernous sinus. The residual tumor within the cavernous sinus was subjected to radiosurgery with tumor growth control during the follow-up period of 2 years. They emphasized the importance of the effectiveness of combined microsurgery and radiosurgery in the treatment of these lesions.

To date, no large radiosurgery series of trigeminal schwannomas has been reported. In our experience, 5 patients showed imaging evidence of decreased tumor size, 6 showed no tumor growth, and 1 showed evidence of tumor progression. Three of the 6 patients with no tumor progression showed imaging evidence of central necrosis. Our prior experience with acoustic schwannomas has shown that tumors showing central necrosis often undergo delayed volume reduction [24].

Unlike microsurgical resection, radiosurgery does not remove a trigeminal schwannoma; instead, the goal is long-term prevention of tumor growth accompanied by low surgical risk. The twin goals of radiosurgery are (1) tumor control and (2) cranial nerve preservation. These goals were achieved in all except 1 of our patients. The only patient in whom treatment failed may represent resistance to radiosurgery of a solitary tumor or tumor growth occurring before the effects of radiosurgery became apparent. In view of the tendency of these tumors to recur after microsurgery [49–51], we believe that residual or recurrent trigeminal schwannomas should be subjected to radiosurgery. Large tumors in elderly or medically infirm patients should be debulked and the residual tumor treated with radiosurgery.

The most frequent complication of radiosurgery is delayed and often transient cranial neuropathies [31]. However, in our experience and in the reports in the literature we found no cranial nerve morbidity following radiosurgery for trigeminal schwannomas. No patient developed radiation-induced edema or other changes in the brainstem during the follow-up period.

The radiobiologic effect of radiosurgery on a schwannoma is probably related to direct tumor cell kill followed by delayed vascular obliteration [58]. In vitro studies suggest that Schwann cells are irreversibly damaged after single-fraction radiation doses as low as 30 Gy. It is believed that tumor cell response is both time- and dose-dependent [24]. Radiosurgery had additional short- and long-term benefits: (1) reduced length of hospital stay (<36 h); (2) reduced patient cost, and (3) rapid return to preoperative functional status (usually within days) [59].

Our results indicate that Gamma knife radiosurgery is a safe primary or adjuvant treatment method for selected patients with trigeminal schwannomas. We anticipate that the long-term rate of growth control will approximate that found with acoustic tumors.

Conclusions

As the long-term results of modern skull base surgery and radiosurgery become available, it appears prudent to follow a cooperative strategy wherein careful preoperative planning aims at microsurgical debulking of difficult skull base lesions with the aim of preserving neural function followed by radiosurgical treatment of the residual lesion. Such a cooperative management strategy planned in advance of the surgical procedure itself might improve the long-term functional status of many patients with complex skull base lesions.

References

1   Harsh GR IV, Joseph MP, Swearingen B, Ojemann R: Anterior midline approaches to the central skull base. Clin Neurosurg 1995;43:15–43.
2   Van Loveren HR, Mahmood A, Liu SS, Gruber D: Innovations in cranial approaches and exposures: Anterolateral approaches. Clin Neurosurg 1995;43:44–52.
3   Pasztor E: The thorny road to minimally invasive techniques in neurosurgery. Minim Invasive Tech Neurosurg 1994;37:64–69.
4   Awad IA: Innovation through minimalism: Assessing emerging technology in neurosurgery. Clin Neurosurg 1995;43:303–316.
5   Samii M, Babu RP, Tatagiba M, et al: Surgical treatment of jugular foramen schwannomas. J Neurosurg 1995;82:924–932.
6   Tan LC, Bordi L, Symon L, et al: Jugular foramen neuromas : A review of 14 cases. Surg Neurol 1990;34:205–211.
7   Clemis JD: Neurogenic tumors of the skull base. Otolaryngol Head Neck Surg 1980;88:511–518.
8   Crumley RL, Wilson C: Schwannomas of the jugular foramen. Laryngoscope. 1984;94:772–777.
9   Fink LH, Early CB, Bryan RN: Glossopharyngeal schwannomas. Surg Neurol 1978;9:239–245.
10 Franklin DJ, Moore GF, Fisch U: Jugular foramen peripheral nerve sheath tumors. Laryngoscope 1989;99:1081–1087.
11 Goldenberg RA, Gardner G: Tumors of the jugular foramen: Surgical preservation of neural function. Otolaryngol Head Neck Surg 1991;104:129.
12 Graham MD, LaRouere MJ, Kartush JM: Jugular foramen schwannomas: Diagnosis and suggestions for surgical management. Skull Base Surg 1991;1:34–38.
13 Hakuba A, Hashi K, Fujitani K, et al: Jugular foramen neurinomas. Surg Neurol 1979;11:83–94.
14 Horn KL, House WF, Hitselberger WE: Schwannomas of the jugular foramen. Laryngoscope 1985;95:761–765.

15   Jackson GC, Cueva RA, Thedinger BA, et al: Cranial nerve preservation in lesions of the jugular fossa. Otolaryngol Head Neck Surg 1991;105:687–693.

16   Kaye AH, Hahn JF, Kinney SM, et al: Jugular foramen schwannomas. J Neurosurg 1984;60: 1045–1053.

17   Kinney SE, Dohn DF, Hahn JF, et al: Neuromas of the jugular foramen; in Brackman DE (ed): Neurological Surgery of the Ear and Skull Base. New York, Raven Press, 1982.

18   Maniglia AJ, Chandler JR, Goodwin WJ Jr, Parker JC Jr: Schwannomas of the parapharyngeal space and jugular foramen. Laryngoscope 1979;89:1405–1414.

19   Raquet F, Mann W, Amedee R, et al: Functional deficits of cranial nerves in patients with jugular foramen lesions. Skull Base Surg 1991;1:117–119.

20   Sasaki T, Takakura K: Twelve cases of jugular foramen neurinoma. Skull Base Surg 1991;1:152–160.

21   Pellet W, Cannoni M, Pech A: The widened transcochlear approach to jugular foramen tumors. J Neurosurg 1988;69:887–894.

22   Pluchino F, Crivelli G, Vaghi MA: Intracranial neurinomas of the nerves of the jugular foramen. Report of 12 personal cases. Acta Neurochir 1975;31:201–221.

23   Flickinger JC, Lunsford LD, Wu A, Maitz A, Kalend AM: Treatment planning for gamma knife radiosurgery with multiple isocenters. Int J Radiol Oncol Biol Phys 1990;18:1495–1501.

24   Linskey ME, Lunsford LD, Flickinger JC: Neuroimaging of acoustic nerve sheath tumors after stereotaxic radiosurgery. AJNR 1991;12:1165–1175.

25   Linskey ME, Lunsford LD, Flickinger JC: Radiosurgery for acoustic neurinomas: Early experience. Neurosurgery 1990;26:736–745.

26   Linskey ME, Lunsford LD, Flickinger JC: Stereotactic radiosurgery for acoustic nerve sheath tumors; in Lunsford LD (ed): Stereotactic Radiosurgery Update. New York, Elsevier, 1991, pp 321–334.

27   Lunsford LD: Contemporary management of meningiomas: Radiation therapy as an adjuvant and radiosurgery as an alternative to surgical removal? J Neurosurg 1994;80:187–190.

28   Lunsford LD, Linskey ME: Stereotactic radiosurgery as an alternative to microsurgery of acoustic tumors. Clin Neurosurg 1992;38:619–634.

29   Lunsford LD, Linskey ME: Stereotactic radiosurgery in the treatment of patients with acoustic tumors. Otolaryngol Clin North Am 1992;25:471–491.

30   Kida Y, Kobayashi T, Tanaka T, Oyama H, Niwa M: Treatment of jugular foramen tumors using radiosurgery. No Shinkei Geka 1995;23:671–675.

31   Pollock BE, Kondziolka D, Flickinger JC, et al: Preservation of cranial nerve function after radiosurgery for nonacoustic schwannomas. Neurosurgery 1993;33:597–601.

32   Dahlin DC, MacCarty CS: Chordoma. A study of fifty-nine cases. Cancer 1952;5:1170–1178.

33   Watkins L, Khudados ES, Kaleoglu M, et al: Skull base chordomas: A review of 38 patients, 1958–1988. Br J Neurosurg 1993;7:241–248.

34   Arnold H, Herman HD: Skull base chordoma with cavernous sinus involvement. Partial or radical removal? Acta Neurochir (Wien) 1986;83:31–37.

35   Handa J, Suzuki F, Nioka H, et al: Clivus chordoma in childhood. Surg Neurol 1987;28:58–62.

36   Heffelfinger HM, Dahlin DC, MacCarty CS, Beabout JW: Chordomas and cartilaginous tumors at the skull base. Cancer 1973;32:410–420.

37   Rich TA, Schiller A, Suit HD, Mankin HJ: Clinical and pathological review of 48 cases of chordoma. Cancer 1985;56:182–187.

38   Bouropoulou V, Bosse A, Roessner A, et al: Immunohistochemical investigation of chordomas: Histogenetic and differential diagnostic aspects. Curr Top Pathol 1989;80:183–203.

39   Krayenbuhl H, Yasargil MG: Cranial chordomas. Prog Neurol Surg 1975;6:380–434.

40   Bourgoiun PM, Tampieri D, Robitaille Y, et al: Low grade myxoid chondrosarcoma of the base of the skull: CT, MR and histopathology. J Comput Assist Tomogr 1992;16:268–273.

41   Al-Mefty A, Borba LA: Skull base chordomas: A management challenge. J Neurosurg 1997;86: 182–189.

42   Gay EL, Sekhar LN, Rubinstein E, et al: Chordomas and chondrosarcomas of the cranial base: Results and follow-up of 60 patients. Neurosurgery 1995;36:887–897.

43   Stapleton SR, Wilkins PR, Archer DJ, Uttley D: Chondrosarcoma of the skull base: A series of eight cases. Neurosurgery 1993;32:348–356.

44 Arseni C, Dumitrescu L, Constantinescu A: Neurinomas of the trigeminal nerve. Surg Neurol 1975; 4:497–503.

45 De Beedittis G, Bernasconi V, Ettoree G: Tumors of the fifth cranial nerve. Acta Neurochir 1977; 38:37–64.

46 Fee We Jr, Epsy CD, Konrad HR: Trigeminal neurinomas. Laryngoscope 1975;85:371–376.

47 Jefferson G: The trigeminal neurinomas with some remarks on malignant invasion of the gasserian ganglion. Clin Neurosurg 1955;1:11–54.

48 Lesoin F, Rosseuax M, Villette A, et al: Neurinomas of the trigeminal nerve. Acta Neurochir 1986; 82:118–122.

49 McCormick PC, Bello JA, Post KD: Trigeminal schwannoma. Surgical series of 14 cases with review of the literature. J Neurosurg 1988;69:850–860.

50 Pollack IF, Sekhar LN, Jannetta PJ, Janecka IP: Neurilemmomas of the trigeminal nerve. J Neurosurg 1989;70:737–745.

51 Samii M, Migliori MM, Tatagiba M, Babu RP: Surgical treatment of trigeminal schwannomas. J Neurosurg 1995;82:711–718.

52 Taha JM, Tew JM Jr, Van Loveren HR, Keller JT, El-Kalliny M: Comparison of conventional and skull base surgical approaches for the excision of trigeminal neurinomas. J Neurosurg 1995;82: 719–725.

53 Yasui T, Hakuba A, Kim SH, Nishimura S: Trigeminal neurinomas: Operative approach in eight cases. J Neurosurg 1989;71:506–511.

54 Schisano G, Olivecrona H: Neurinomas of the Gasserian ganglion and trigeminal root. J Neurosurg 1960;17:306–322.

55 Dolenc VV: Frontotemporal epidural approach to trigeminal neurinomas. Acta Neurochir 1994; 130:55–65.

56 Kida Y, Kobayashi T, Tanaka T: Gamma radiosurgery of intracranial neuromas other than acoustic tumors. Neuroradiology 1995;37:599.

57 Yamasaki T, Nagao S, Kagawa T, Takamura M, Moritake K, Tanaka T, Kida Y, Kobayashi T: Therapeutic effectiveness of combined microsurgery and radiosurgery in a patient with a huge trigeminal neurinoma (Japanese). No to Shinkei (Brain Nerve) 1996;48:845–850.

58 Anniko M, Arndt J, Noren G: The human acoustic neuroma in organ culture. II. Tissue changes after gamma irradiation. Acta Otolaryngol (Stockh) 1981;91:223–235.

59 Lunsford LD: Contemporary management of meningiomas: Radiation therapy as an adjuvant and radiosurgery as an alternative to surgical removal? J Neurosurg 1994;80:187–190.

N. Muthukumar, MCh, FACS, Department of Neurosurgery,
Madurai Medical College and Government Rajaji Hospital, Madurai (India)
Fax: +91 (452) 531 056

Lunsford LD, Kondziolka D, Flickinger JC (eds): Gamma Knife Brain Surgery.
Prog Neurol Surg. Basel, Karger, 1998, vol 14, pp 145–159

Radiosurgery Management of Brain Metastasis from Systemic Cancer

John C. Flickinger, Douglas Kondziolka, L. Dade Lunsford

Departments of Radiation Oncology, Neurological Surgery and Radiology, and
the Center for Image-Guided Neurosurgery, University of Pittsburgh,
Pittsburgh, Pa., USA

Extent of the Problem

Brain metastases develop in 10–30% of patients with cancer [18]. The proportion of cancer patients detected who will be found to harbor a brain metastasis will most likely increase in the future. Improved systemic therapy (chemotherapy and immunotherapy) should lengthen the survival of patients with metastatic cancer, allowing more time for metastases to develop in the brain. Most chemotherapy or immunotherapy drugs have difficulty penetrating the blood-brain barrier; this makes the central nervous system a sanctuary site for metastatic cancer. The recent development of immunotherapy regimens compound the problem of even asymptomatic brain metastases. Because these therapies can increase intracranial peritumoral edema, screening to detect subclinical brain metastases is necessary. This results in metastatic being detected earlier in the brain at smaller sizes. Furthermore, these subclinical metastases must be treated aggressively with surgery or radiosurgery before immunotherapy can either start or continue.

Pickren et al. [18] analyzed the distribution of brain metastases in an autopsy series of 10,916 cancer patients at Roswell Park Memorial Institute between 1959 and 1979. The incidence of brain metastases detected at autopsy in this series was 8.7% (67% had systemic metastases with no brain metastases and 24% had no metastatic disease). The three cancer diagnoses with the highest rates of brain metastasis at autopsy were: skin (28%), lung (23%), and metastatic carcinoma with unknown primary site (21%).

Figure 1 shows the distribution of the number of brain metastases detected at autopsy. Solitary tumors were found in 39% of the cases with brain meta-

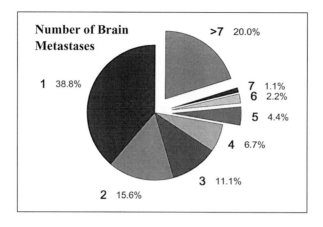

Fig. 1. Distribution of the number of brain metastases found at autopsy in 954 patients with brain metastasis [18].

stasis, two or less in 54% and four or fewer brain metastases in 72% of cases. Eight or more metastases were identified in only 20% of the patients. This study supports the concept that brain metastases occur in limited numbers and are therefore amenable to a local therapy such as surgery or radiosurgery.

Unlike surgical resection, which is difficult to perform with more than one brain metastasis, the number of metastases treatable by radiosurgery with an efficient tool (such as a Gamma knife or a dedicated modified Linac) is limited only by the patience of the radiosurgery team. There is no clear consensus on how many metastases are treatable. From 1987 to 1989 we selected only patients with one tumor for radiosurgery. We then decided to allow entry of patients with only 2–4 metastases into our multiple brain metastases trial that randomizes patients initially treated by 30 Gy of whole-brain irradiation (XRT) to up-front radiosurgery versus observation (with radiosurgery reserved for post-XRT tumor progression). We are somewhat flexible with our upper limit of four metastases for radiosurgery in patients whose tumors progress following whole-brain XRT, and occasionally treat up to six depending upon the size and number of metastases, and also the expected survival if brain metastases can be controlled (depending on the extent of systemic disease, functional status, age, etc.)

The brain was the only site of cancer found at autopsy in 3.4% of these patients with brain metastases in the Roswell series. It is in this small group of patients with no metastatic disease outside the brain that aggressive regional therapy to the brain should have the greatest impact on survival. In the other 97% of brain metastasis patients with extracranial metastases, effective brain metastasis management can lead to improved survival and quality of life. The survival benefit conferred by effective management of brain metastases in patients with metastatic disease outside the brain depends on how well the extracranial disease is controlled.

Whole-Brain Radiotherapy

One of the first questions investigated by the Radiation Therapy Oncology Group (RTOG) was what fractionation schedule for whole-brain XRT worked best for brain metastases [10]. They assessed five different radiation fractionation schedules for whole-brain XRT in 1,830 brain metastasis patients: 20 Gy in 5 fractions, 30 Gy in 10 or 15 fractions, and 40 Gy in 15 or 20 fractions. Analysis of subsequent survival could not detect any difference among the five different fractionation regimens. Even though 20 Gy in five fractions appeared equally effective, most radiotherapists in the United States treat brain metastasis patients with whole-brain XRT to 30 Gy in 10 or 12 fractions because of a concern that large fractions will lead to more delayed neurological deterioration. In the RTOG study, ambulatory patients with no systemic metastases had the greatest median survival (28 weeks) compared to nonambulatory patients (11 weeks).

The extent to which brain metastases eventually regrow after conventional radiotherapy has not always been fully appreciated by radiotherapists. Many did not maintain long-term follow-up with these patients who were usually closely managed by medical oncologists and because patients with recurrent brain metastases were not referred back for repeat radiotherapy. Patchell et al. [17] demonstrated the extent of this problem by reporting an actuarial recurrence rate exceeding 80% for whole-brain radiotherapy (36 Gy in 12 fractions) of solitary brain metastasis in his study, where it compared unfavorably to surgery. For this reason, conventional fractionated whole-brain radiotherapy alone may not be adequate treatment in good prognosis patients (ambulatory with no other systemic metastases).

Recent information on late sequelae after whole-brain XRT provides another reason to be more concerned with its use in managing brain metastases in patients with good prognoses. The routine use of prophylactic cranial irradiation (PCI) in patients with small cell lung cancer is presently controversial because of late effects on brain. It is easier to interpret follow-up studies of neurological function in PCI patients than in patients who start out with problems from clinically apparent brain metastases. Fleck et al. [6] reported significant neurological toxicity in 7 of 11 small cell lung cancer patients surviving more than 30 months after PCI to 30 Gy in 10 fractions. Lishner et al. [12] looked at 58 small cell lung cancer patients who survived more than 24 months out of a total series of 641 treated and came to a different conclusion.

They documented delayed nonmetastatic neurological complications in 9/48 patients (19%) who received PCI to 20 Gy and an identical percentage of 20% in 2/10 2-year survivors without PCI. Although this series points out the fact that not all late neurological deterioration in these patients is from whole-brain XRT, there is still reason to be concerned about the late effects of whole-brain XRT in patients who are expected to be long-term survivors from treatment of their brain metastasis.

Surgery of Brain Metastases

Because of the poor long-term control of brain metastasis with whole-brain XRT alone, resection of solitary brain metastasis has been advocated in good prognosis patients [3, 15, 16, 23]. Randomized trials by Patchell et al. [17] and Noordijk et al. [16] demonstrated not only superior tumor control, but also superior survival for patients undergoing surgical resection plus radiotherapy compared to radiotherapy alone (median survivals: 40 vs. 15 weeks for the Patchell study and 10 vs. 6 months for the Noordijk study). These studies indicate that the outcomes after conventional radiotherapy alone are worse than after surgical resection followed by whole-brain XRT.

Unfortunately, surgical resection is not an option for patients with deeply located tumors or high medical risks for surgery. Furthermore, the benefit of surgical resection may have been overstated in the older literature. A recent larger randomized study by Mintz et al. [15] with 84 patients failed to show any benefit for surgical resection followed by whole-brain XRT compared to whole-brain radiotherapy alone in either survival (median 5.6 months of S + XRT vs. 6.3 months of XRT alone) or functionally independent survival.

We feel that patients who have solitary brain metastasis with symptoms from mass effect, not relieved by steroids, benefit from surgical resection when they are acceptable surgical candidates (no medical contraindications, limited systemic disease). All other patients are better candidates for either whole-brain XRT and/or radiosurgery.

Radiosurgery Results

In general, the survival and morbidity results of radiosurgery are equal or superior to those reported for surgical resection followed by whole-brain XRT. Table 1 lists several large representative series of patients treated by radiosurgery for brain metastases, and shows radiosurgery to be a treatment modality with a high local control rate and low complication rate in comparison to surgical

Table 1. Results of radiosurgery for brain metastases in representative series

Institution	Patients with metastases	D_{min}, Gy (median)	% local control	% necrosis	Median survival, months
University of Pittsburgh					
lung cancer	115/77	16	85	4	10
melanoma	118/60	16	90	0	7
renal cell	52/35	17	90	0	11
Harvard JCRT [1]	421/248	15	83 [65 [a]]	3 [7 [a]]	9
UCSF [21]	261/119	18.5	93 [77 [a]]	4	11
Karolinska [11]	235/160	30	94	NS [b]	NS [b]
Heidelberg [5]	102/69	17	94	3	6 or 12 [c]
Auchter: multi-institutional [2]	122/122	17	86 [77 [a]]	0	12
Gamma Knife Users Group [8]	116/116	16	85 [67 [a]]	4	11
Stanford [9]	52/33	25	94	4	NS [b]
University of Wisconsin [14]	58/40	18	82	0	6.5

[a] Actuarial.

[b] NS, not stated.

[c] Seven months with multiple metastases and 12 months with solitary metastases.

resection. Any comparison of these impressive results with surgical resection must also consider the fact that location is not an exclusionary criteria for radiosurgery and that patients who are poor candidates for surgery for other reasons (poor medical condition, etc.) are also included in these series.

The Gamma Knife Users Group Study

The Gamma Knife Users Group studied the results of radiosurgery in 116 patients with solitary brain metastasis [8]. The median follow-up after radiosurgery was 7 months, while the median follow-up after the diagnosis of brain metastasis was 12 months. Radiosurgery was part of the initial brain metastasis management of 71 patients, while 45 patients had tumors that regrew after prior whole-brain XRT. Minimum radiosurgery tumor doses varied from 8 to 30 Gy (mean 17.5 Gy). Fifty-one patients received radiosurgery alone (usually for recurrent disease) and 65 patients underwent whole-brain radiotherapy (mean dose 34 Gy) combined with radiosurgery (within 2 months of each other). As shown in table 1, the median survival was 11 months from radiosurgery and 20 months after diagnosis. The median survival following diagnosis of brain metastasis was 43 months for patients with recurrent tumors versus 14 months for patients receiving radiosurgery at presentation. This is a good example of lead time bias, since patients with recurrent tumors have already survived up to the point of tumor recurrence. Multivariate analysis of postradiosurgery survival found that histology was the only significant independent predictive variable ($p=0.041$). Survival was better for breast cancer (just as in the RTOG whole-brain XRT study) and worse for melanoma and renal cell cancer compared to other histologies.

The actuarial rate of local tumor control in the Gamma Knife Users Group Study was 67% at 2 years. Univariate and multivariate analysis identified significantly improved tumor control in patients that received radiosurgery and whole-brain radiotherapy compared to radiosurgery alone (81 vs. 53% at 2 year, $p=0.011$). Patients who received radiosurgery alone in this series predominantly had recurrent tumors and received a similar dose (median $D_{min}=17$ Gy) as patients who had radiosurgery combined with whole-brain XRT (median $D_{min}=16$ Gy). Since other groups have reported higher tumor control rates with radiosurgery alone to higher doses, it appears that the improvement in local tumor control from adding whole-brain radiotherapy seen in the Gamma Knife Users Group Study can also be achieved by increasing the dose used for radiosurgery alone [11, 21, 22].

Another finding in the Gamma Knife Users Group Study was that patients with melanoma or renal cell carcinoma had better local control compared to other histologies ($p=0.0006$). These patients actually had a poorer survival than those with other histologies. It is possible that the high rate of death from systemic disease in these patients resulted in an overestimation of the local control rate and that the true local control rate is not any different than the other tumors. The 2-year actuarial rate for development of radiation necrosis requiring surgical resection was 4% and for symptomatic edema requiring steroid therapy was 11%. The 2-year actuarial rate of tumor hemorrhage after radiosurgery was 8%.

The Harvard JCRT Experience

Alexander et al. [1] reported the Harvard Joint Center experience with 421 brain metastases in 248 patients treated by Linac radiosurgery from 1986 to 1993. Median follow-up was 36 months. The number of brain metastases treated in each patient were: one metastasis in 177, two metastases in 52, and three or more metastases in 19 patients. Radiosurgery was used as part of initial brain metastasis management along with fractionated whole-brain XRT in 60 patients, while 188 patients underwent radiosurgery for recurrent tumors. Seventy-seven patients had no known systemic extracranial metastases, while another 171 patients had 'stable' extracranial metastases. In 30% of the patients, relatively radioresistant brain metastases (melanoma, renal cell carcinoma or sarcoma) were treated. The median minimum tumor dose from radiosurgery (D_{min}) was 15 Gy. The median target volume was 3 cm^3.

Multivariate analysis identified significantly better survival than survival with the lack of any active systemic disease (either no other metastases or stable disease) and age less than 60. The actuarial tumor control rates in this study were similar to the Gamma Knife Users Group Study (85% at 1 year and 65% at 2 years). Increased local failure was significantly associated with recurrent tumors and infratentorial location in multivariate analysis. They found a higher rate of local failure rate with infratentorial tumors. Inadequate imaging of the posterior fossa appears to be the most likely explanation, since CT scans may show bone artifacts leading to inadequate target definition. The addition or substitution of MRI should eliminate this problem in the future. We have seen no difference in tumor control for posterior fossa tumors in the experience at the University of Pittsburgh, where MR is routinely used for targeting. Complications in the Harvard JCRT series were limited to 7% of patients with mass effect requiring surgical resection and 1% of patients developing cranial neuropathies.

The UCSF Series

Shiau et al. [21] recently reported the experience of the University of California, San Francisco (UCSF) with brain metastasis radiosurgery. They studied the responses of 219 brain metastases in 100 patients. The median volume treated was 1.3 cm^3 (range 0.02–31 cm^3) and the median minimum target dose was 18.5 Gy (range 10–22 Gy). They reported a 77% 1-year actuarial rate for freedom from progression for all tumors and 90% for 90 lesions that had a radiosurgery dose prescribed of 18 Gy or more. Multivariate analysis identified significantly better freedom from progression for metastases with a homogeneous pattern of contrast enhancement (present before radiosurgery in 68% of tumors), higher prescribed radiosurgery dose (more significant than combined biologically equivalent doses of radiosurgery plus fractionated whole-brain XRT) and for a longer interval primary diagnosis and radiosurgery. They could not identify any independent significant differences in freedom from progression among patients treated at presentation versus recurrence or with/without whole-brain XRT.

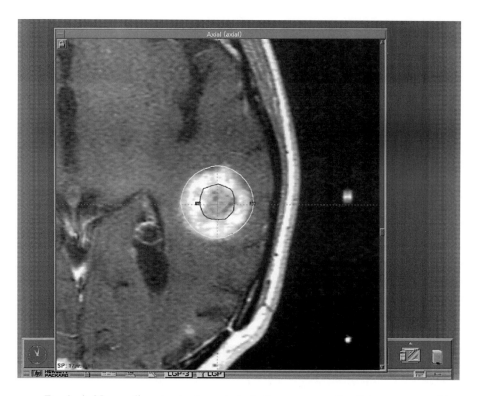

Fig. 2. A 25-mm diameter brain metastasis from non-small cell lung cancer with a central low-density region that fails to take up contrast (left side). A radiosurgery treatment plan is superimposed (right side). The tumor is enclosed by the 30% isodose volume while the 90% isodose encloses the majority of the low-density (nonenhancing) region 90%.

We subsequently verified the UCSF finding of poorer local control with radiosurgery for inhomogeneously enhancing brain metastases in the University of Pittsburgh patients. The relative radioresistance observed for brain metastases with inhomogeneous contrast enhancement might be from a greater hypoxic tumor cell population within these tumors. Figure 2 shows a novel way to overcome radioresistance of these probably hypoxic zones with intratumoral boost radiosurgery. The plan shown encloses the entire tumor within the 30% isodose with the low-density region that fails to take up contrast enclosed within the 90% isodose. The higher dose within the hypoxic portion of the tumor should make up for any radioresistance since the oxygen enhancement ratio for fully hypoxic tumor should be no greater than 3.

University of Pittsburgh Series Stratified by Histology

We recently reviewed our experience with brain metastasis radiosurgery in separate analyses for each of the most common histologies treated (table 1). The three most common brain metastases treated by gamma knife radiosurgery at the University of Pittsburgh were non-small cell lung cancer, malignant melanoma, and renal cell carcinoma.

Our review of radiosurgery for non-small cell lung cancer brain metastases looked at 77 patients with 115 tumors. Median tumor volume was 3.2 cm^3 (range 0.03–21.8 cm^3). The median margin tumor dose was 16 Gy (range 10–22.5 Gy). Seventy-one patients had whole-brain XRT. Thirty-four patients were treated for tumor recurrence after whole-brain XRT; 37 had combined XRT and radiosurgery as initial management. The overall median survival was 10 months from radiosurgery and 15 months from diagnosis of brain metastasis. Multivariate analysis correlated significantly improved survival with lack of active systemic disease, tumor diameter <2 cm, lack of intra-tumoral necrosis, and resection of the primary lung cancer. Local tumor control was maintained in 77/91 evaluable metastases (85%) and 88% of the patients treated. Multivariate analysis of local tumor control identified a significant improvement with small tumor volume (<3 cm^3). Only 3 patients underwent craniotomy and tumor resection for tumor growth. One patient underwent resection for tumoral necrosis and 3 for subsequent intratumoral hemorrhage. Follow-up imaging identified development of peritumoral edema in 12 patients.

Our recent study of melanoma brain metastases included 60 consecutive patients with 118 tumors treated by radiosurgery. Thirty-six patients (60%) had solitary brain metastases. The mean volume was 2.95 cm^3 (range 0.1–25.5 cm^3). The mean margin tumor dose was 16.4 Gy (range, 10–20 Gy). Fifty-one patients had whole-brain XRT (15 as initial treatment before re-growth, 36 as combined initial therapy with radiosurgery). Median survival was 7 months from radiosurgery and 10 months from brain metastasis diagnosis. In patients with a solitary tumor and no other active systemic disease, median survival was 15 months. The local tumor control rate was 90% (70/77) with local failure only in tumors receiving ≤16 Gy. Remote brain metastases developed in 14 patients. We found no significant difference in the incidence of remote brain metastasis development in patients treated with or without whole-brain XRT (44 vs. 23%, p=0.97).

Our recent review of renal cell carcinoma metastases included 35 patients with 52 brain metastases managed with radiosurgery. The mean tumor volume was 2.4 cm^3 (range 0.1–14.1 cm^3). The mean marginal tumor dose was 17 Gy (range 13–20 Gy). Median survival was 11 months postradiosurgery and 14 months from brain metastasis diagnosis. Multivariate analysis correlated improved survival with age < 55, no active systemic disease, and use of systemic therapy after radiosurgery. Local tumor control was 90% (35/39 evaluable tumors). Remote brain metastases developed in 12 patients. Whole-brain XRT did not seem to lower the rate of distant brain metastasis development. Two patients developed delayed symptomatic radiation sequelae requiring steroids. No patient suffered a postradiosurgery hemorrhage.

Brain Tolerance for Radiosurgery of Metastases

As shown in table 1, radiation necrosis has not been much of a problem in most radiosurgery series for brain metastasis. Shaw et al. [20] recently reported the RTOG dose-escalation study for the radiosurgery of recurrent brain metastases or primary tumors. Unacceptable toxicity for dose escalation was defined as irreversible RTOG grade 3, or any grade 4–5, CNS toxicity in > 20% of patients per treatment arm within 3 months of radiosurgery. The responses of 156 patients (64% with metastasis) were analyzed. There were 31 patients treated at two academic gamma knife facilities (Mayo Clinic and UCSF) and 125 patients treated at a number of different Linac facilities. Serious quality control problems were found such as radiosurgery treatment volumes that were more than a factor of 2 larger than the target or smaller than the target. Escalations of minimum tumor dose (D_{min}) were stratified by tumor diameter. For tumors < 20 mm in diameter, D_{min} of 18, 21 and 24 Gy were tested; 0/40 patients developed grade 3 or greater CNS toxicity. For tumors 21–30 mm in diameter, D_{min} of 15, 18 and 21 Gy were tested. Two out of 42 of these patients developed grade 3 CNS toxicity at dose levels of 15 and 21 Gy. Patients with 31–40 mm diameter tumors developed more serious toxicity problems. At the dose level of 12 Gy, 1/21 patients died (grade 5 toxicity). At 15 Gy, 0/22 patients had grade 3 or greater toxicity, but at 18 Gy, 4/18 patients developed significant toxicity (2 with grade 3 and 2 with grade 4). The maximum tolerated dose for 3–4 cm diameter tumors was therefore judged to be 18 Gy.

We do not feel that the RTOG dose-escalation study adequately addresses long-term toxicity of radiosurgery. We usually do not exceed the traditional dose-volume guidelines from the integrated logistic formula or the Kjellberg 1% isoeffect line, because of the adequate tumor control and limited toxicity reported in series using these guidelines [1, 7, 8].

Treatment Strategies

There are several different strategies that can be used to manage patients with brain metastases. Patients with small solitary brain metastases are excellent candidates for either radiosurgery alone or combined with whole-brain XRT, if they have a good prognosis (no known extracranial systemic disease, good functional status). Patients with brain metastasis from relatively radioresistant tumors like malignant melanoma may not benefit from adding whole-brain XRT to radiosurgery [1, 8, 21, 24, 25]. In these patients, whole-brain XRT, with resulting hair loss and possible late effects on cognitive function, could be withheld until other brain metastases develop; at which time further radiosurgery and/or whole-brain XRT could be administered.

Past policies of whole-brain XRT for all patients with brain metastasis need to be reconsidered in light of the limited effectiveness of whole-brain XRT, better ability to image subclinical (small asymptomatic) metastases, documented late effects of whole-brain XRT in long-term survivors, and better ability to localize and treat subsequently developing metastases without overlapping with prior radiation treatment volumes.

The guidelines are unclear for when radiosurgery should be used in patients with solitary brain metastasis at presentation (with or without whole-brain XRT) versus treating with whole-brain XRT alone initially and reserving radiosurgery for salvage therapy. Most neurosurgeons and radiation oncologists would agree that the latter approach of withholding aggressive local therapy like radiosurgery or surgical resection is too conservative for young patients with no other systemic disease. Such an approach may be appropriate in older patients with poor functional status and systemic disease progressing in the face of systemic therapy [1, 7, 15]. We feel that since radiosurgery has proven effectiveness and low morbidity, radiosurgery must be proven to be not worthwhile in well-defined patient groups before we can in good conscience relegate patients to therapy that appears less effective.

Using radiosurgery as part of initial management of patients with multiple metastases is another controversial issue. The conventional wisdom of the past that more than one brain metastasis is a herald of tumor seeding throughout the brain is contradicted by autopsy studies and the high local control for patients with two, three and even more metastases with both surgical resection and radiosurgery [1, 18, 21]. The vast majority of these patients seem to develop only limited numbers of brain metastases over time [18]. We have been trying to address this question by a phase III clinical trial for patients with 2–4 metastases. All receive whole-brain XRT to 30 Gy/12 fractions and then are randomized to either immediate radiosurgery or a plan to use radiosurgery only as a salvage treatment for post-XRT tumor progression.

Surgery Compared with Radiosurgery

Rutigliano et al. [19] published an extensive cost-benefit comparison of gamma knife radiosurgery and surgical resection for solitary brain metastasis. The study concluded that radiosurgery had a lower uncomplicated procedure cost, a lower average complications cost per case, a lower total cost per procedure, was more effective, and had a better incremental cost effectiveness per life-year. Treatment-related morbidity and mortality were higher for surgery (30 and 7% respectively) than for radiosurgery (4 and 0%). The radiosurgery data for the comparison was based primarily on the Gamma Knife Users Group Study which was the only published radiosurgery study at the time with only solitary metastasis treated at presentation in the majority of patients that could be compared to surgical resection series.

A thorough comparison of the effectiveness of surgical resection and radiosurgery was recently completed by Auchter et al. [2], who compiled a multi-institutional series of patients who underwent radiosurgery as part of initial management for *resectable* solitary brain metastasis to compare with surgical series. They presented a strong case for replacing surgical resection as the treatment of choice for small solitary brain metastasis. Their series of patients with resectable brain metastasis appears comparable to the patients in the randomized trials of Patchell et al. [17] and Noordijk et al. [16] that showed the superiority of surgical resection followed by whole-brain radiotherapy over whole-brain XRT alone. Auchter's series included 122 patients who met their entry criteria (solitary metastasis by CT or MR with documented extracranial cancer, surgically resectable metastasis, no prior brain XRT or brain surgery, histology that is not small cell, lymphoma or germ cell, no urgent need for surgery, Karnofsky ≥ 70, and age ≥ 18). These 122 patients were all that met the entry criteria out of 533 patients who received Linac radiosurgery at the participating institutions. Treatment volumes were 0.13–27.2 cm^3 with a median of 2.7 cm^3. D_{min} varied from 10 to 27 Gy with a median of 17 Gy. Whole-brain XRT doses were 25–40 Gy with a median of 37.5 Gy. Median follow-up was 123 weeks.

The overall initial response rate was 59% (25% complete, 34% partial). Stable disease was seen in 36% and progressive disease in 6%. In-field recurrence developed in 14% and intracranial out-of-radiosurgery-field relapse in 22%. Median survival was 56 weeks with 25% of the deaths from CNS progression. Multivariate analysis identified the presence of extracranial metastases and poor Karnofsky score to be related to decreased survival (but not age or histology). Table 2 compares median overall survival and survival with KPS ≥ 70 (independent function) for the whole-brain XRT and surgery groups from the randomized trials of Patchell et al. [17] and Noordijk et al. [16]. We

Table 2. Comparison of median overall survival and functionally independent survival (KPS functional score ≥70) for patients with resectable solitary brain metastasis treated in the whole-brain XRT and surgical resection plus XRT arms from the randomized trials of Patchell et al. [17], Noordijk et al. [16] and Mintz et al. [15] to the matched series of patients undergoing radiosurgery in the multi-institutional series of Auchter et al. [2]

Series	n	Treatment arm	Median overall survival, weeks	Median survival with KPS ≥ 70, weeks
Auchter	122	Radiosurgery	56	44
Patchell	25	Surgery + whole-brain XRT	40	33
Noordijk	32	Surgery + whole-brain XRT	43	38
Mintz	41	Surgery + whole-brain XRT	24	32% of all days
Patchell	23	whole-brain XRT	15	8
Noordijk	31	whole-brain XRT	26	15
Mintz	43	whole-brain XRT	27	32% of all days

have also added the data for the more recent study by Mintz et al. [15] which was not included in Auchter's comparison. Both overall and independent median survivals were better than the whole-brain XRT and surgery arms of the three randomized studies. The actuarial local failure rate was lower for the radiosurgery study (14%) compared to the surgery arm (20%) and whole-brain XRT arm (52%) of Patchell's study; actuarial local control was not reported in the other two series. The large number of patients (n = 122) in Auchter's multi-institutional radiosurgery study lends weight to the comparison with the smaller studies by Patchell (n = 48), Noordijk (n = 63), and Mintz (n = 84).

Between Auchter's series and the other published results of radiosurgery, which also include unresectable brain metastases, it is safe to conclude that radiosurgery is *at least* equally effective for most patients. By virtue of its lower morbidity and cost, it is the superior choice for most patients [2, 13, 19]. Surgical resection should be reserved for patients with large resectable tumors (> 3 cm in diameter) with associated mass effect and significant neurological deterioration.

The best initial treatment for every different patient with brain metastasis remains to be defined. Current options include fractionated radiotherapy alone, surgery plus radiation therapy, surgery alone, radiosurgery alone or radiosurgery plus radiation therapy. So far, the latter seems to be the standard for effective treatment of brain metastasis with low morbidity.

References

1   Alexander E III, Moriarty TM, Davis RB, Wen PY, Fine HA, Black PM, Kooy HM, Loeffler JS:
    Stereotactic radiosurgery for the definitive, noninvasive treatment of brain metastases. J Natl Cancer
    Inst 1995;87:34–40.
2   Auchter RM, Lamond JP, Alexander EA, Buatti JM, Chappell R, Friedman WA, Kinsella TJ, Levin
    AB, Noyes WR, Schultz CJ, Loeffler JS, Mehta MP: A multi-institutional outcome and prognostic
    factor analysis of radiosurgery for resectable single brain metastasis. Int J Radiat Oncol Biol Phys
    1996;35:27–36.
3   Bindal RK, Sawaya R, Leavens ME, et al: Surgical treatment of multiple brain metastases. J Neurosurg
    1993;79:210–216.
4   Borgelt B, Gelber R, Kramer S, Brady L, Chang C, Davis L, Perez C, Hendrickson F: The palliation
    of brain metastases: The final results of the first two studies by the radiation therapy oncology group.
    Int J Radiat Oncol Biol Phys 1980;6:1–9.
5   Engenhart R, Kimmig BN, Hover KH, et al: Long-term follow-up for brain metastases treated by
    percutaneous stereotactic single high-dose irradiation. Cancer 1993;71:1353–1361.
6   Fleck JF, Einhorn LH, Lauer RC, Schultz SM, Miller ME: Is prophylactic cranial irradiation indicated
    in small-cell lung cancer? J Clin Oncol 1990;8:209–214.
7   Flickinger JC: The integrated logistic formula and prediction of complications from radiosurgery. Int
    J Radiat Oncol Biol Phys 1989;17:879.
8   Flickinger JC, Kondziolka D, Lunsford LD, et al: A multi-institutional experience with stereotactic
    radiosurgery for solitary brain metastasis. Int J Radiat Oncol Biol Phys 1994;28:797–802.
9   Fuller BG, Kaplan ID, Adler J, et al: Stereotaxic radiosurgery for brain metastases: The importance
    of adjuvant whole-brain irradiation. Int J Radiat Oncol Biol Phys 1992;23:413–418.
10 Gelber R, Larson M, Borget BB, Kramer S: Equivalence of radiation schedules for the palliative
    treatment of brain metastases in patients with favorable prognosis. Cancer 1981;48:1749–1753.
11 Kihlstrom L, Karlsson B, Lindquist C: Gamma knife surgery for cerebral metastasis: Implications for
    survival based on 16 years' experience. Stereotact Funct Neurosurg 1993;61(supp 1):45–50.
12 Lishner M, Feld R, Payne DG, et al: Late neurological complications after prophylactic cranial
    irradiation in patients with small-cell lung cancer: The Toronto experience. J Clin Oncol 1990;8:215–
    221.
13 Loeffler JS, Shrieve DC: What is appropriate therapy for a patient with a single brain metastasis? Int
    J Radiat Oncol Biol Phys 1994;29:915–917.
14 Mehta MP, Rozental JM, Levin AB, et al: Defining the role of radiosurgery in the management of
    brain metastases. Int J Radiat Oncol Biol Phys 1992;24:619–625.
15 Mintz AH, Kestle J, Rathbone MP, et al: A randomized trial to assess the efficacy of surgery in addition
    to radiotherapy in patients with a single cerebral metastasis. Cancer 1996;78:1470–1476.
16 Noordijk EM, Vecht CJ, Haaxma-Reiche H, et al: The choice of treatment of single brain metastasis
    should be based on extracranial tumor activity and age. Int J Radiat Oncol Biol Phys 1994;29:
    711–717.
17 Patchell RA, Tibbs PA, Walsh JW, et al: A randomized trial of surgery in the treatment of single
    metastases to the brain. N Engl J Med 1990;322:494.
18 Pickren JW, Lopez G, Tzukada Y, Lane WW: Brain metastases. An autopsy study. Cancer Treat Symp
    1983;2:295–313.
19 Rutigliano MJ, Lunsford LD, Kondziolka D, Strauss MJ, Khanna V, Green M: The cost effectiveness of
    stereotactic radiosurgery versus surgical resection in the treatment of solitary metastatic brain tumors.
    Neurosurgery 1995;37:445–455.
20 Shaw E, Farnan N, Souhami L, Dinapoli R, Kline R, Loeffler J, Fisher B: Radiosurgical treatment of
    previously irradiated primary brain tumors and brain metastasis: Final report of Radiation Oncology
    Group (RTOG) protocol 90-05. Int J Radiat Oncol Biol Phys 1996;34:647–654.
21 Shiau CY, Sneed PK, Shu HKG, et al: Radiosurgery for brain metastases: Relationship of dose and
    pattern of enhancement to local tumor control. Int J Radiat Oncol Biol Phys 1997;37:385–391.
22 Shirato H, Takamura A, Tomita M, et al: Stereotactic irradiation without whole-brain irradiation for
    single brain metastasis. Int J Radiat Oncol Biol Phys 1997;37:385–391.

23   Smalley SR, Laws ER, O'Fallon JR, Shaw EG, Schray MF: Resection for solitary brain metastasis:
     Role of adjuvant radiation and prognostic variables in 229 patients. J Neurosurg 1992;77:531–540.
24   Somaza S, Kondziolka D, Lunsford LD, Kirkwood J, Flickinger J: Stereotactic radiosurgery for cere-
     bral metastatic melanoma. J Neurosurg 1993;79:661–666.
25   Vlock DR, Kirkwood JM, Leutzinger C, Kapp DS, Fischer JJ: High-dose fraction radiation therapy
     for intracranial metastases of malignant melanoma. Cancer 1982;49:2289–2294.

John C. Flickinger, MD, University of Pittsburgh Medical Center,
Department of Radiation Oncology, 200 Lothrop Street, Pittsburgh, PA 15213 (USA)
Tel. (412) 647 3600, Fax (412) 647 6029

Lunsford LD, Kondziolka D, Flickinger JC (eds): Gamma Knife Brain Surgery.
Prog Neurol Surg. Basel, Karger, 1998, vol 14, pp 160–174

....

Stereotactic Radiosurgery for Glial Neoplasms

Douglas Kondziolka, John C. Flickinger, L. Dade Lunsford

Department of Neurological Surgery, and Radiation Oncology,
Center for Image-Guided Neurosurgery, University of Pittsburgh, Pa., USA

Glial neoplasms of the brain are associated with variable outcomes. Multimodality management including surgery, radiation therapy, boost irradiation techniques and chemotherapy appear to enhance survival and quality of life in appropriate patients [8, 10, 21, 27]. Despite improvements in each of these treatments, most patients eventually develop local tumor progression which leads to neurologic morbidity and death. Factors that influence outcome include histologic grade, tumor type, patient age, brain location, radiation dose, Karnofsky performance status, and surgical resectability [5, 34]. In 1987 we began to use stereotactic radiosurgery as a less invasive method to boost conventional tumor irradiation, or replace radiation therapy for selected benign tumors [7]. Initial criticisms of the use of radiosurgery for gliomas included its focal radiation delivery [14] and its delivery during a single session. However, the intense radiobiologic effect of single-session radiation cell kill, regardless of mitotic phase, was our argument for the use of radiosurgery.

Radiobiology of Radiosurgery for Glial Tumors

During radiosurgery a focused volume of radiation is delivered to an intracranial target in a single treatment session. For benign tumors, optimal results are achieved when the targeted tumor volume matches precisely the radiosurgery volume. In the case of a malignant glial tumor, this can never be achieved. Although the contrast-enhanced tumor volume identified on imaging can be irradiated by a conformal margin isodose during radiosurgery, the malignant tumor cells beyond that identified by contrast enhancement, remain 'outside' the radiosurgery volume [18]. This of course makes malignant glioma radiosurgery different from AVM and benign tumor radiosurgery, or even perhaps metastatic tumor radiosurgery with a comparatively smaller zone of regional infiltration. This is not to say that these peripheral malignant cells receive *no* irradiation during radiosurgery, but that they exist in the fall-off of radiation outside the selected isodose. Nevertheless, the performance of a focused surgical procedure to the volume of contrast enhancement (whether through craniotomy and tumor resection, brachytherapy, or radiosurgery), can provide potential benefits to the patient by improving the likelihood of local control [9, 13, 15, 35]. Since treatment failure usually occurs locally [3], a local radiation boost might be expected to provide some benefit. For this reason, several groups have initiated malignant glioma radiosurgery programs [22, 25, 26]. Barker et al. [2] reported a large 1988 to 1993 experience using fractionated radiation therapy (60 Gy) alone for adjuvant management of glioblastoma multiforme. Median survival in this series was 11.2 months with a 2-year survival rate of 16%. This modern comprehensive series of results provides a concurrent historical experience to compare the potential benefits of boost radiosurgery.

The radiobiology of radiosurgery is different from that of brachytherapy. In radiosurgery, radiation is delivered in a single session, usually over 10–60 min depending on dose and dose rate. The biologic effect of radiation delivery in this single session is high, and the goal is to arrest cell-division capability irrespective of an individual cell's mitotic phase during irradiation. In brachy-therapy (interstitial irradiation), a focal radiation dose is delivered over 4–6 days and thus an attempt is made to exploit susceptibility of cells within the cell cycle [1, 12, 20, 23]. Shrieve et al. [31] compared radiosurgery (n = 72) with brachytherapy (n = 32) in patients with glioblastoma. The mean tumor volume at radiosurgery was 10.1 cm^3 and the mean margin dose 13 Gy. One third of their patients had received prior chemotherapy. Reoperation after radiosurgery was performed in 22% of patients as compared to 44% of patients in the brachytherapy group. The histologic findings in these patients showed both residual tumor and tumor necrosis. Median survival after radiosurgery for recurrent tumors was 10.2 months and the 2-year survival rate 19%.

There has been no laboratory comparison using in vivo tumor models. However, we compared the effects of 35-Gy radiosurgery on the rat C6 glioma and a 10-fraction, 85-Gy fractionated dose calculated to be of biological equivalence [19]. In this controlled study, animal survival and reduction in tumor size were identical. However, more pronounced histologic effects were seen in the radiosurgery group when a high maximum dose was delivered. Cytotoxic effects indicated an early and more direct effect of radiation in this malignant tumor model. In contrast, indirect vascular obliterative effects are

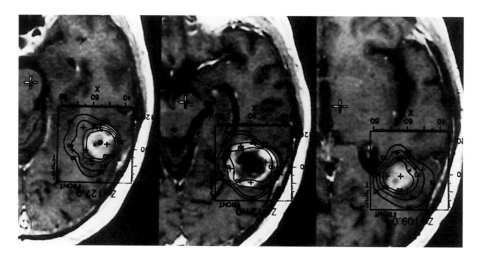

Fig. 1. Axial MR images showing the radiosurgery dose plan to manage a 68-year-old man with a recurrent posterior left temporal lobe glioblastoma. A six isocenter plan was targeted to the contrast-enhanced margin (50% isodose line) to deliver a margin dose of 16 Gy and a maximum dose of 32 Gy.

seen in benign tumor radiosurgery models. Thus, in an experimental model, radiosurgery was found to provide a positive clinical and histologic benefit for a malignant glial tumor [19]. We believe that radiosurgery is appropriate as part of a management plan for patients with glioblastoma multiforme or anaplastic astrocytoma.

Technique of Radiosurgery

In patients with glial tumors, we performed stereotactic radiosurgery under local anesthesia except children younger than 14 years of age who had radiosurgery under general anesthesia [11]. Between 1987 and 1991, patients had stereotactic computed tomography (CT) imaging for tumor localization. Patients managed since 1991 underwent stereotactic magnetic resonance imaging (MRI) for tumor definition. The edge of the contrast-enhanced portion of the mass was used to identify the tumor volume for radiosurgical targeting (fig. 1). For glial tumors that did not exhibit contrast enhancement, we performed radiosurgery to the volume of long TR (relaxation time) signal change. The isodose configuration was made such that the selected treatment isodose enclosed the margin of contrast enhancement. Tumor cells beyond this rim of peripheral enhancement (within the low density region seen on CT scan or the high signal region seen on long TR MRI) were radiated at a dose below the peripheral selected isodose in the region of radiation fall-off. A single

40-mg intravenous dose of methylprednisolone was administered at the end of radiation. Initial imaging follow-up was performed 6–8 weeks after radiosurgery and then at 3- to 6-month intervals thereafter.

Malignant Gliomas

Gamma knife radiosurgery was performed in 113 glioblastoma multiforme patients and 50 anaplastic astrocytoma patients using the following entry criteria: mean tumor contrast-enhanced tumor diameter <3.5 cm, age <75 years, Karnofsky performance status ≥50, any brain location, and proven histologic diagnosis. This series comprised 7% of the total number of patients undergoing stereotactic radiosurgery at the University of Pittsburgh during our initial 10-year experience. We used radiosurgery to treat the contrast-enhancing portion of the tumor, and not tumor infiltration beyond that border at the same dose. The fall-off in radiation dose occured into infiltrated regional brain.

Glioblastoma multiforme

For initial histologic diagnosis, 73 patients underwent craniotomy and resection (65%) and 40 had a stereotactic biopsy. Then mean patient age was 50 years (range 3–83) and the mean Karnofsky performance score was 90 (range 50–100) at the time of radiosurgery. Radiosurgery was part of an initial management strategy that included either postdiagnosis conventional fractionated radiation therapy followed by radiosurgery, or three cycles of continuous infusion intravenous BCNU/cisplatinum chemotherapy followed by radiation therapy and radiosurgery. In this instance radiosurgery was typically performed 5–8 months postdiagnosis. The mean fractionated radiation therapy (tumor plus 3 cm) dose was 60 Gy.

The mean tumor volume (calculated by the dose-volume histogram) at radiosurgery was 4.8 ml (range 0.88–31.2). The 50% isodose line was used to cover the tumor margin in 90% of patients (range 40–90% isodose). The mean dose delivered to the tumor margin was 15 Gy (range 12–25) and the mean maximum dose 30.2 Gy (range 21.4–52). The tumors were located in hemispheric (lobar) locations in 104 patients (92%) and deep (posterior fossa or diencephalon) locations in 9 patients.

We performed a detailed analysis of the first 64 patients managed with radiosurgery [17]. At the time of analysis, 34 patients had died and 30 were alive. Median survival after radiosurgery for the entire group was 16 months (range 1–74) [4, 16, 24] (table 1). The median survival after initial diagnosis was 26 months (range 5–108) (fig. 2). When radiosurgery was performed at

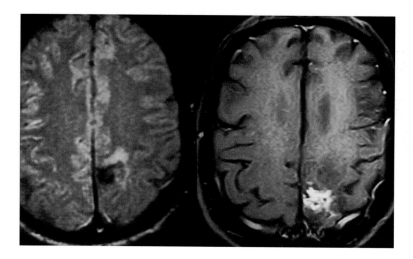

Fig. 2. Left: MR scan at the time of radiosurgery in a 54-year-old man with a glioblastoma of the left frontal-parietal region. He received a tumor margin dose of 15 Gy. *Right:* Nine years after radiosurgery and fractionated radiotherapy, no tumor progression was identified.

Table 1. Survival after malignant glioma radiosurgery

Timing	Histology	Median survial months	Two-year survival, %	Three-year survival, %
Postdiagnosis	GBM	26	51	30
Postradiosurgery	GBM	16	38	21
Postdiagnosis	AA	32	67	40
Postradiosurgery	AA	21	49	32

GBM = Glioblastoma multiforme; AA = anaplastic astrocytoma.

the time of tumor progression, median survival afterwards was 30 months (range 2–74). The 2-year survival rate from diagnosis for the overall series was 51%. When radiosurgery was performed as part of initial therapy in 45 patients, the median survival post-diagnosis was 20 months (range 5–76). Two-year survival was 41% [4, 16, 24]. No survival benefit was identified for patients who had intravenous chemotherapy in addition to radiosurgery.

After radiosurgery, 12 patients (19%) required delayed craniotomy and resection and 4 patients (6%) underwent a second radiosurgery for either local progression or progression at a separate location. The mean time to craniotomy after radiosurgery was 5 months or to a second radiosurgery, 7 months. In 11 of 12 patients who underwent craniotomy after radiosurgery and radiation therapy, histologic findings included a mixture of viable tumor and necrosis with radiation effect. In 1 patient only total necrosis without tumor was found.

Anaplasic Astrocytoma

The mean patient age of this group was 37 years (range 3–73) and the mean Karnofsky performance score was 90 (range 50–100). The histologic diagnosis of anaplastic astrocytoma was obtained during craniotomy and resection in 20 patients (40%) and by stereotactic biopsy in 30 patients. The tumor was located in a lobar location in 31 patients (62%) and in a deep location in 19 patients.

Stereotactic radiosurgery was performed at a mean of 4 months for the initial treatment group and 20 months for the disease progression group. The mean tumor volume at radiosurgery was 4.6 ml (range 18–20.1 ml). The mean dose delivered to the tumor margin was 15.6 Gy (range 10–20 Gy); the mean maximum tumor dose was 30.6 Gy (range 18.7–36 Gy).

We performed a detailed analysis in the first 43 patients [17]. Twenty-three patients were alive and 20 had died. Median survival after radiosurgery was 21 months (range 3–93 months). The median survival after tumor diagnosis was 32 months (range 5–96). For the 23 patients who had radiosurgery at the time of disease progression, median survival after radiosurgery was 31 months (range 3–47). The 2-year survival rate after diagnosis was 67%. No survival difference was found in patients who had intravenous chemotherapy. For the 21 patients who had radiosurgery as part of the intial treatment program, median survival was 56 months postdiagnosis (range 9–93). The 2-year survival rate was 88%.

After radiosurgery, 10 patients (23%) underwent craniotomy and resection, a mean of 8 months after radiosurgery. Pathologic findings included a mixture of viable tumor and radiation changes in 8 patients and total tumoral radiation necrosis in 2. Two patients (5%) underwent a second-staged radiosurgery procedure for regional tumor progression. No patient suffered acute neurologic morbidity after radiosurgery and no patient suffered seizures after treatment. Two patients with anaplastic astrocytomas (4.7%) developed new, delayed neurologic morbidity related to radiosurgery. Both had imaging evidence of an adverse radiation effect and recovered. One of these patients required sugical resection.

Table 2. Multivariate analysis of factors related to survival after glioblastoma (GBM) or anaplastic astrocytoma (AA) radiosurgery

Factor	Improved postdiagnosis survival (p value)	Improved postradiosurgery survival (p value)
Younger age[1]	0.0001	0.0008
AA vs. GBM	0.03	0.19
Biopsy vs. craniotomy	0.41	0.26
Smaller tumor volume[1]	0.17	0.02
Margin dose ≥ 16 Gy	0.64	0.50
Lobar location	0.95	0.98
Karnofsky score ≥ 70	0.004	0.001

[1] Continuous variable.

Comparison with the Radiation Therapy Oncology Group (RTOG) Recursive Partitioning Analysis

Table 2 lists the results of multivariate analysis of variables suspected of influencing survival of the combined series of 107 patients. Table 3 shows the survival of the 65 patients in this series who had radiosurgery as part of initial therapy divided into RTOG recursive partitioning classes [6]. This technique is valuable for comparing series data using clinical and histologic stratification parameters. Neither RTOG class nor extent of surgery (needle biopsy versus resection) proved to be significant predictors of survival. Age proved to be the factor most predictive of survival either from diagnosis or from radiosurgery. Table 3 compares the median and 2-year survivals of 65 patients treated at presentation with radiosurgery in this series with 1,578 patients treated without radiosurgery in three RTOG trials stratified by RTOG class. This suggests that patients in RTOG class III–V benefit from radiosurgery and patients with class I–II (anaplastic astrocytoma with age < 50 and normal mental status or age ≥ 50 and symptoms for > 3 months) do not. However, the patient population in this study is much more selective than the RTOG study since patients in this series all had limited disease suitable for radiosurgery, and 38% were selected only after they developed new tumor progression. Second, it is possible that the poor prognosis patients in the RTOG classes derive a proportionally greater benefit from radiosurgery than the better RTOG classes. A recursive partitioning classification system that separates patients into more than two age groups might be better able to predict prognosis because of the overwhelming prognostic importance of this factor. At this time we believe that the designation of patients with small volume tumors as eligible or ineligible for radiosurgery with current available data is premature.

Table 3. University of Pittsburgh versus RTOG survivals in patients with glioblastoma (GBM) or anaplastic astrocytoma (AA)

RTOG grade	University of Pittsburgh			RTOG		
	n	median survival months	2-year survival, %	n	median survival months	2-year survival, %
I	13	37	67	139	59	76
II	2	5	50	34	37	68
III	13	39	73	175	18	35
IV	11	16	24	457	11	15
V	24	19	26	395	9	6
VI	2	7	0	263	5	4

RTOG Class: I AA, normal mention, age <50.
II AA, age >50, KPS 70–100, symptoms >3 months.
III Age <50, AA+abnormal mentation; GBM, KPS ≥90.
IV GBM, age <50, KPS <90; age >50, KPS >70, AA <3 months or GBM employed after resection.
V Age >50, KPS ≥70, GBM postresection, XRT >54 Gy, not employed; or KPS <70 with normal mentation.
VI Age >50, KPS <70, abnormal mentation, or postresection but received <54 Gy XRT.

Other Reports and Effect on Survival

To date, there has not been a randomized trial to study the effects of radiosurgery, nor has there been even a large single-center review. Masciopinto et al. [25] reported 31 patients who had radiosurgery for glioblastoma multiforme using a linear accelerator; median survival was 9.5 months. The mean tumor volume as 16.4 cm^3, significantly larger than the 6.5-ml mean volume observed in this study. All patients had radiation therapy to 50–66 Gy followed by radiosurgery. Regional recurrence occurred in 65% of patients at a mean of 7 months. They concluded from their data that radiosurgery provided no survival benefit compared to surgery plus fractionated radiation therapy alone. They believed that this lack of benefit was attributed to larger tumor volumes [25].

On the other hand, Sarkaria et al. [28] reported improved results after surgery, fractionated radiation therapy, and radiosurgery in the initial management of 115 patients with either glioblastoma multiforme or analplastic astrocytoma. The tumor margin dose varied between 10 and 20 Gy. The median survival for glioblastoma was 91 weeks and the 2-year survival rate 38% (41% in the present series). For anaplastic astrocytoma, median survival was not reached in this series and the 2-year survival rate was 72%, similar to our 67% rate. Disease progression was observed in 59% of patients and reoperation performed in 29% (again similar to our findings). Stea et al. [33] found no difference in survival between 33 brachytherapy and 19 radiosurgery patients. However, in this limited series a lower toxicity rate after radiosurgery may have been due to a relatively low median radiosurgery dose of 10 Gy.

To study reoperation after linear accelerator radiosurgery, Schwartz et al. [29] reported a median survival from diagnosis of glioblastoma of 18 months and after radiosurgery, 11 months. For the group of patients who underwent reoperation after radiosurgery, median survival was increased to 24 months postdiagnosis, and 16 months postradiosurgery [28]. Thirty-one percent of their patients required surgery after radiosurgery at a mean time of 5.5 months. Although this reoperative rate was slightly higher than in our series, we attribute this mainly to the treatment of larger tumors.

Radiosurgery for Tumor Progression

When clinical and imaging evidence of tumor progression is documented, management options for many patients are limited. When a lobar tumor progresses, craniotomy and resection (debulking) is often possible. When an infiltrative malignant neoplasm is progressing, is a focused therapy such as radiosurgery appropriate [30]? Because most tumor growth is due to local progression, we hypothesized that a specific radiosurgical boost could provide a significant survival benefit for malignant glioma patients. When radiosurgery was used at the time of progression, we found that glioblastoma patients survived a median of 30 months. Anaplastic astrocytoma patients survived a median of 31 months. As a well-tolerated, low-morbidity procedure with a less than 24-hour hospital stay, we believe that radiosurgery provided a high level of expected palliation with minimal detrimental effect on quality of life. Thus, for selected patients with progressive small-volume malignant glial tumors after prior therapy, radiosurgery potentially provides safe and effective palliation with enhanced local growth control. McDermott et al. [26] from the University of California at San Francisco, reported on the use of radiosurgery for recurrent glioblastoma multiforme (n = 34) and anaplastic astrocytoma (n = 12). Median survival for glioblastoma was 40.6 weeks and for anaplastic astrocytoma 61.6 weeks.

Pilocytic Astrocytoma

Pilocytic astrocytoma is a distinct histologic subtype of astrocytoma that is most often diagnosed in children and young adults [32]. Pilocytic astrocytomas are usually well circumsribed on CT and MRI and are strongly contrast-enhancing. Many tumors have an indolent course and long-term survival (and even cure) after complete surgical resection. However, tumor recurrence or progression may occur after subtotal resection in a high-risk brain location. The effectiveness of fractionated external beam radiation therapy has been questioned. The fact that pilocytic astrocytomas can be cured by resection indicates that it should have sharp margins that make it potentially curable with focused irradiation. We anticipated that single-fraction, high-dose volumetric irradiation of these often well-circumscribed tumors might provide a superior alternative to other therapies [32].

To date, 23 patients with pilocytic astrocytoma had stereotactic radiosurgery at our center. Mean patient age was 16 years (range 4–58). The mean duration of symptoms prior to radiosurgery was 2 years. Fifteen patients had undergone prior resection and 15 had a neurologic deficit at the time of radiosurgery. Four patients were treated for tumor progression after prior fractionated radiation therapy. Tumor locations included the cerebellum (n = 7), thalamus (n = 3), pons/midbrain (n = 6), frontal lobe (n = 2), temporal lobe (n = 3), and intraventricular (n = 1). Mean tumor volume at radiosurgery was 1.5 ml (range 0.06–7.1). The mean dose delivered to the tumor margin was 15.2 Gy (range 13–20) and a mean maximum dose of 30 Gy (range 21.4–40). After radiosurgery, 21 of 23 patients had stabilization of their tumor volume (fig. 3). Two patients developed tumor enlargement due to radiation effect with subsequent regression. In a detailed review of the first 9 children with longer-term follow-up, 5 tumors had decreased in size and 4 showed no further growth [32]. No child developed a new or progressive neurologic deficit from tumor growth. Six children had improvement in their preoperative neurologic deficit, and in the first 9 patients, school performance and intellectual/emotional development were judged to be unchanged by family members after radiosurgery [32].

We believe that radiosurgery is valuable for pilocytic astrocytomas because these tumors are often well circumscribed histologically. Radiosurgery likely leads to tumor growth arrest by preventing further cellular division coupled with delayed intratumoral vascular obliteration. Many patients with pilocytic astrocytomas (even after resection) remain clinically and biologically dormant without any other intervention. However, children with progressive, well-circumscribed tumors that are located in critical or deep areas are potential candidates for radiosurgery. Larger tumors (>3.5 cm in diameter) should again be considered for resection. Fortunately, many brainstem pilocytic astrocytomas are small and suitable for radiosurgery. Long-term outcomes remain to be identified.

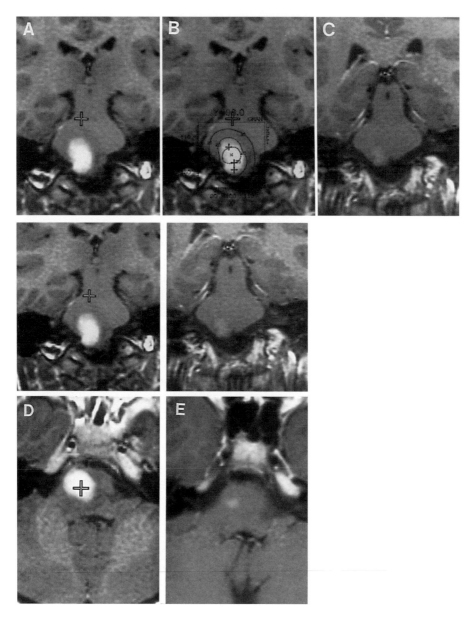

Fig. 3. MR images in a child with a pilocytic astrocytoma of the pons residual after resection (*A, D*). Radiosurgery was performed using three isocenters (*B*). One year later, significant tumor regression was found (*C, E*). Four years after radiosurgery she remains in excellent neurologic condition with no imaging evidence of tumor.

Astrocytoma and Oligodendroglioma

Patients with nonpilocytic, nonanaplastic astrocytomas may be candidates for stereotactic radiosurgery. To date, 15 patients were managed at our institution at a mean age of 25 years (range 3–57). Six patients had prior external beam radiation therapy. Subtotal resection had been performed in 4 patients. Brain locations included frontal lobe (n = 2), temporal lobe (n = 2), hypothalamus (n = 1), occipital lobe (n = 1), intraventricular (n = 4), thalamus (n = 1) and brainstem (n = 4). The mean tumor volume was 2.5 ml and the mean tumor margin dose 15.5 Gy. In follow-up, 2 patients had imaging disappearance of their tumor, 6 had a decrease in tumor size, 5 were unchanged, and 1 increased (mean follow-up of 3 years, maximum 7). For an astrocytoma to be suitable for radiosurgical targeting, it should be of small volume and in a location where larger field fractionated radiation might wish to be avoided [11]. For well-circumscribed histologically proven astrocytomas, radiosurgery should be considered as potentially less toxic and perhaps of more therapeutic benefit than fractionated radiotherapy. We await data on long-term outcomes.

Ten patients with oligodendrogliomas had radiosurgery. Six had received prior radiation therapy and 7 had undergone prior subtotal resection. Thus, radiosurgery was used as a strategy for recurrent tumor in most patients, with similar treatment parameters as noted above for astrocytoma. Although there was no morbidity from radiosurgery, 2 patients had further tumor growth. Eight patients continue to be followed, now at a maximum of 4 years.

Ependymoma

Radiosurgery has been used as an adjuvant management for recurrent or residual ependymomas after prior microsurgical resection [11, 22]. Twenty-six patients had radiosurgery (mean age of 25 years) at our center. Eighteen patients (69%) had undergone prior external beam radiation therapy before radiosurgery. Twenty-one patients had undergone prior resection (maximum four resections). Tumor locations included fourth ventricle and brainstem (n = 15), cerebellum (n = 3), middle fossa skull base (n = 2), frontal lobe (n = 1), temporal lobe (n = 1), parietal lobe (n = 1) and cavernous sinus (n = 1) (fig. 4). The mean tumor volume was 4.3 ml (range 0.21–26 ml). Mean dose delivered to the tumor margin was 14 Gy (range 10–20) and the mean maximum dose 28.7 Gy. Impressive local tumor responses were noted in many patients. Those with anaplastic ependymomas often recurred at a distance, even when initial regression was identified.

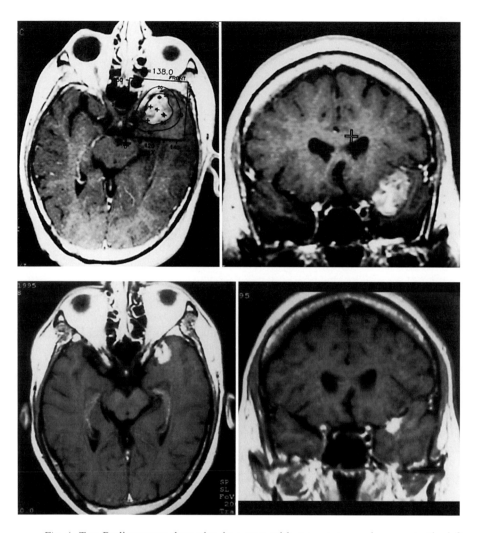

Fig. 4. Top: Radiosurgery dose plan in a man with recurrent ependymoma to the left Sylvian fissure region. He had prior resection and radiation therapy for a fourth ventricular ependymoma. *Bottom*: Six months following radiosurgery, significant tumor regression was noted.

The role of radiosurgery for ependymoma continues to be defined. When the biologic behavior of the tumor enters a phase of rapid progression, radiosurgery can be an effective palliation for circumscribed small volume tumors, but the tendency for these tumors to recur outside their imaging-defined borders remains problematic. We have observed dramatic reduction of irradiated tumors only to identify new tumor growth beyond the radiosurgery margin. Since most of these patients have already failed large-feld fractionated radiotherapy, further treatment options are limited. Effective therapy may await the development of tumor-specific radiation sensitizers or brain-protectant agents that permit the delivery of a higher radiosurgical dose.

References

1 Agbi CB, Bernstein M, Laperriere N, Leung P, Lumley M: Paterns of recurrence of malignant astrocytoma following stereotactic interstitial brachytherapy with iodine-125 implants. Int J Radiat Oncol Biol Phys 1992;23:321–326.

2 Barder F, Prados MD, Chang SM, Gutin P, Lamborn K, Larson D, Malec MK, McDermott M, Sneed PK, Wara W, Wilson CB: Radiation response and survival time in patients with glioblastoma multiforme. J Neurosurg 1996;84:442–448.

3 Bashir R, Hochberg F, Oot R: Regrowth patterns of glioblastoma multiforme related to planning of interstitial brachytherapy radiation fields. Neurosurgery 1988;23:27–30.

4 Brookmeyer R, Crowley J: A confidence interval for the median survival time. Biometrics 1982;38: 29–41.

5 Coffey RJ, Lunsford DL, Taylor FH: Survival after stereotactic biopsy of malignant gliomas. Neurosurgery 1988;22:465–473.

6 Curran WJ, Scott CB, Horton J, Nelson J, Winstein A, Fischbach J, Chang C, Rotman M, Asbell S, Krisch R, Nelson D: Recursive partitioning analysis of prognostic factors in three Radiation Therapy Oncology Group malignant glioma trials. J Natl Cancer Inst 1993;85:704–710.

7 Dempsey RK, Kondziolka D, Lunsford LD, Flickinger JC: The role of stereotactic radiosurgery in the treatment of glial tumors; in Lunsford LD (ed): Stereotactic Radiosurgery Update. Amsterdam, Elsevier Science, 1992, pp 407–410.

8 Deutsch M, Green SB, Strike TA, et al: Results of a randomized trial comparing BCNU plus radiotherapy, streptozotocin plus radiotherapy, BCNU plus hyperfractionated radiotherapy, and BCNU following misonidazole plus radiotherapy in the postoperative treatment of malignant glioma. Int J Radiat Oncol Biol Phys 1989;16:1389–1396.

9 Davaux BC, O'Fallon JR, Kelly PJ: Resection, biopsy, and survival in malignant glial neoplasmas. A retrospective study of clinical parameters, therapy, and outcome. J Neurosurg 1993;78:767–775.

10 Florell RC, Macdonald DR, Irish WD, Bernstein M, Leibel S, Gutin P, Cairncross JG: Selection bias, survival, and brachytherapy for glioma. J Neurosurg 1992;76:179–182.

11 Grabb P, Lunsford LD, Albright AL, Kondziolka D, Flickinger JC: Stereotactic radiosurgery for glial neoplasms of children. Neurosurgery 1996;38:696–702.

12 Green SB, Shapiro WR, Burger PC, et al: A randomized trial of interstitial radiotherapy boost for newly diagnosed malignant glioma: Brain Tumor Cooperative Group Trial 8701. Proc Am Soc Clin Oncol 1994;13:174.

13 Gutin PH, Leibel SA, Wara WM, Choucair A, Levin VA, Phillips T, Silver P, Da Silva V, Edwards MSB, Davis RL, Weaver KA, Lamb S: Recurrent malignant gliomas: Survival following interstitial brachytherapy with high-activity iodine-125 sources. J Neurosurg 1987;67:864–873.

14 Halperin EC, Burger PC, Bullard DE: The fallacy of the localized supratentorial malignant glioma. Int J Radiat Oncol Biol Phys 1988;15:505–509.

15 Harsh GR, Levin VA, Gutin PH, Seager M, Silver P, Wilson CB: Re-operation for recurrent glioblastoma and anaplastic astrocytoma. Neurosurgery 1987;21:615–621.

16 Kaplan EL, Meier P: Nonparametric estimation from incomplete observations. J Am Stat Assoc 1958;53:457–481.

17 Kondziolka D, Flickinger JC, Bissonette D, Bozik M, Lunsford LD: The survival benefit of stereotactic radiosurgery for patients with malignant glial neoplasms. Neurosurgery 1997;41:776–785.

18   Kondziolka D, Lunsford LD, Claassen D, Pandalai S, Maitz A, Flickinger JC: Radiobiology of radiosurgery. II. The rat C6 glioma model. Neurosurgery 1992;31:280–288.
19   Kondziolka D, Somaza S, Comey C, Lunsford LD, Classen D, Pandalai S, Maitz A, Flickinger JC: Radiosurgery and fractionated radiation therapy: Comparison of techniques in an in vivo rat glioma model. J Neurosurg 1996;84:1033–1038.
20   Laperriere NJ: Critical appraisal of experimental radiation modalities for malignant astrocytomas. Can J Neurol Sci 1990;17:199–208.
21   Leibel SA, Sheline GE: Radiation therapy for neoplasms of the brain. J Neurosurg 1987;66:1–22.
22   Loeffler JS, Alexander E III, Shea WM, Wen P, Fine HA, Kooy H, Black PM: Radiosurgery as part of the initial management of patients with malignant gliomas. J Clin Oncol 1992;10:1379–1385.
23   Loeffler JS, Alexander E III, Wen PY, Shea W, Coleman CN, Kooy H, Fine H, Nedz L, Silver B, Riese N, Black PM: Results of stereotactic brachytherapy used in the initial management of patients with glioblastoma. J Natl Cancer Inst 1990;82:1918–1921.
24   Mantel N: Evaluation of survival data and two new rank order statistics arising in its consideration. Cancer Chemother Rep 1966;50:163–170.
25   Masciopinto JE, Levin AB, Mehta MP, Rhode BS: Stereotactic radiosurgery for glioblastoma: A final report of 31 patients. J Neurosurg 1995;82:530–535.
26   McDermott MW, Sneed PK, Chang SM, Gutin P, Wara W, Verhey L, Smith V, Petti P, Ho M, Park E, Edwards MSB, Prados MD, Larson DA: Results of radiosurgery for recurrent gliomas; in Kondziolka D (ed): Radiosurgery 1995. Radiosurgery. Basel, Karger, 1996, vol 1, pp 102–112.
27   Nazzaro JM, Neuwelt EA: The role of surgery in the management of supratentorial intermediate and high-grade astrocytomas in adults. J Neurosurg 1990;73:331–344.
28   Sarkaria JN, Mehta MP, Loeffler JS, Buatti J, Chappell RJ, Leven A, Alexander E, Friedman W, Kinsella T: Radiosurgery in the initial management of malignant gliomas: Survival comparison with the RTOG recursive partitioning analysis. Int J Radiat Oncol Biol Phys 1995;32:931–941.
29   Schwartz MS, Loeffler JS, Black PM, Shrieve D, Wen PY, Fine HA, Alexander E: Reoperation following radiosurgery of glioblastoma: Impact on survival and neurologic status; in Kondziolka D (ed): Radiosurgery 1995. Radiosurgery. Basel, Karger, 1996, vol 1, pp 141–157.
30   Shaw E, Scott C, Souhami L, Dinapoli R, Bahary JP, Kline R, Wharam M, Schultz C, Davey P, Loeffler JS, Del Rowe J, Marks L, Fisher B, Shin K: Radiosurgery for the treatment of previously irradiated recurrent primary brain tumors and brain metastases: Initial report of Radiation Therapy Oncology Group protocol 90-05. Int J Radiat Oncol Biol Phys 1996;34:647–654.
31   Shrieve DC, Alexander E III, Wen PY, Fine HA, Kooy H, Black PM, Loeffler JS: Comparison of stereotactic radiosurgery and brachytherapy in the treatment of recurrent glioblastoma multiforme. Neurosurgery 1995;36:275–284.
32   Somaza S, Kondziolka D, Lunsford LD, Flickinger JC, Bissonette D, Albright AL: Early outcomes after stereotactic radiosurgery for growing pilocytic astrocytomas in children. Pediatr Neurosurg 1996;25:109–115.
33   Stea B, Rossman K, Kittelson J, Lulu B, Shetter A, Cassady JR, Hamilton A: A comparison of survival between radiosurgery and stereotactic implants for malignant astrocytomas. Acta Neurochir 1994;62(suppl):47–54.
34   Walker MD, Strike TA, Sheline GE: An analysis of dose-effect relationship in the radiotherapy of malignant gliomas. Int J Radiat Oncol Biol Phys 1979;5:1725–1731.
35   Wilson CB: Glioblastoma: The past, the present, and the future. Clin Neurosurg 1992;38:32–48.

Douglas Kondziolka, MD, University of Pittsburgh Medical Center, Suite B-400,
Department of Neurological Surgery, 200 Lothrop Street, Pittsburgh, PA 15213 (USA)
Tel. (412) 647 6782, Fax (412) 647 0989

Lunsford LD, Kondziolka D, Flickinger JC (eds): Gamma Knife Brain Surgery.
Prog Neurol Surg. Basel, Karger, 1998, vol 14, pp 175–194

....

Stereotactic Radiosurgery in the Treatment of Pineal Region Tumors

Brian R. Subach[a], *L. Dade Lunsford*[a–c], *Douglas Kondziolka*[a,b]

Departments of [a]Neurological Surgery, [b]Radiation Oncology, [c]Radiology, and
the Center for Image-Guided Neurosurgery, University of Pittsburgh
Medical Center, Pittsburgh, Pa., USA

The optimal management of lesions arising from the pineal region has long been the subject of debate. The tumors arising in this area represent the entire spectrum of neoplasia, with cells of origin including glial, meningeal, pineal parenchymal, and germ cell with behaviors ranging from benign to highly malignant. The signs and symptoms attributed to these masses generally result from aqueductal obstruction or tectal plate compression. Most frequently, children and young adults, those in the second and third decades of life, are affected. The onset of symptoms may be rapid or insidious with common presentations including headache, mental status changes, papilledema, or gaze paresis [6, 8, 25]. Many will have obstructive hydrocephalus and require cerebrospinal fluid (CSF) diversion as an initial procedure.

Evaluation of a pineal region mass requires a combination of neurodiagnostic imaging, tumor marker assessment, and biopsy. Initial magnetic resonance imaging (MRI) should consist of unenhanced and enhanced multiplanar views of the brain and spine to define anatomic relationships and stage the disease. Additional information may be gained by examining blood and CSF for cytology and specific markers [32, 50]. The tumors in this region exhibit a variable propensity to disseminate along ependymal surfaces and throughout the subarachnoid space. Demonstration of positive cytology will alter the need for surgical intervention and for adjunctive therapy [62]. Similarly, serum and CSF testing for germ cell markers, including α-fetoprotein (AFP), the β-subunit of human chorionic gonadotropin (β-HCG) and placental alkaline phosphatase (PLAP) may yield additional diagnostic information. In most cases, the diagnosis is not clearly established until diagnostic tissue is obtained. With the advanced operative techniques available today, some advocate stereotactic biopsy, while others use open biopsy for tissue diagnosis prior to initiation of definitive treatment.

Diagnostic and Treatment Strategies

The number of treatments developed over the years for tumors of the pineal region are rivaled only by the number of different tumor subtypes arising from this area. Such heterogeneity, in the face of the relative rarity of such tumors, limits the ability to define a single effective management strategy. The initial results of open surgical intervention were accompanied by significant morbidity. Given the anatomic complexity of the area and the presence of critical brain and vascular structures, early reports of mortality associated with resection ranged from 50 to 70% [17, 34, 66]. As a consequence of the dismal outcomes, many surgeons considered these tumors unresectable and opted for conservative management. During the 1950s–1970s, mass lesions in the pineal region were often treated by empiric fractionated radiation therapy without histopathologic confirmation of tumor identity [19, 25]. After administration of a 'test dose' (e.g. 20 Gy), the patient would undergo neuroimaging to assess effectiveness of initial treatment and to determine the need for further radiation versus surgical intervention. Tumor shrinkage was considered 'diagnostic' of germinoma. Observation of stable tumor size led to completion of the standard fractionated radiation dose, while tumor growth mandated surgery.

In the 1970s, advances in microsurgical techniques led many surgeons to advocate open biopsy or resection as a means to obtain both a definitive diagnosis and treatment for the tumor. Although often successful in identifying the neoplasm, the likelihood of curative resection for most tumors remained relatively low [10, 15, 21, 32, 38, 54]. Similarly, a number of patients with radiosensitive or highly malignant tumors were subjected unnecessarily to the risks of craniotomy. Both groups required a second treatment modality, thereby calling into question the value of open surgery.

The combined evolution of stereotactic technology and neurodiagnostic imaging has led to the widespreaed use of image-guided, minimal-access techniques to diagnose and treat a variety of intracranial lesions (fig. 1). The application of such techniques to the problem of pineal region tumors has resulted in a safe and effective alternative to open biopsy, resection, and empiric fractionated radiation therapy. The widely varied spectrum of neoplasms arising from this region makes the possibility of a single treatment regimen for all tumors implausible. In cases where the diagnosis is not made by typical

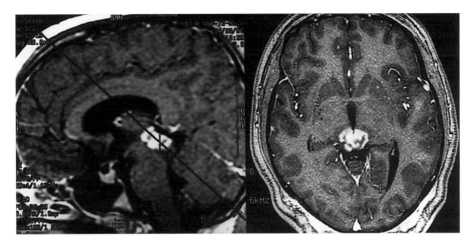

Fig. 1. Sagittal and axial MRI scans demonstrating pineal region germinoma and typical biopsy trajectory.

neuroimaging criteria, the use of stereotactic biopsy allows formation of a rational treatment plan designed to maximize the benefits of resection, fractionated radiation therapy, chemotherapy, or stereotactic radiosurgery [15, 22, 42, 52, 56]. Patients with benign tumors such as meningiomas and mature teratomas are often candidates for operative intervention. Malignant tumors such as gliomas and metastases may be best treated by a combination of fractionated radiation therapy and radiosurgery [16]. There is recent evidence to suggest that tumors such as pineocytoma, pineoblastoma and germ cell neoplasms may be best treated either by stereotactic radiosurgery alone or in combination with adjunctive chemotherapy or radiation therapy [5, 7, 11, 23].

Pathologic Considerations

Pineal region neoplasms may be divided into three subgroups: (1) pineal parenchymal tumors (pineocytoma and pineoblastoma); (2) germ cell tumors (germinoma and nongerminomatous germ cell subtypes) and (3) other tumors arising from glial, meningeal, lymphoid, and connective tissues.

Pineal Parenchymal Tumors

Tumors arising from the parenchyma of the pineal gland have been subdivided into three subtypes: pineocytoma (pinealoma), pineoblastoma and a mixed tumor subtype. All are exceedingly rare and represent a minority of pediatric and adult brain tumors [9, 13, 45]. The pineocytoma is a low-grade tumor which arises from the pineal cells. Often difficult to distinguish from

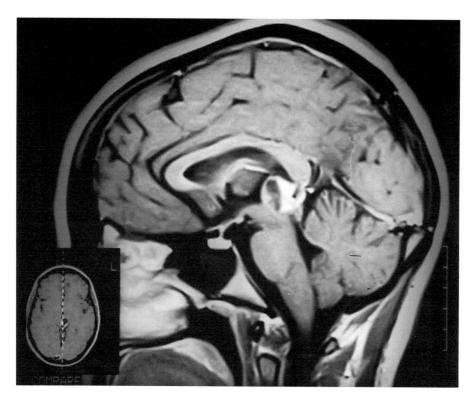

Fig. 2. Well-demarcated lesion with inhomogeneous contrast enhancement typical of a pineocytoma.

normal glandular elements, this tumor commonly presents in young adults. It is typically slow growing with few anaplastic features [50]. These tumors frequently calcify, remain well demarcated from surrounding brain, and inhomogeneously enhance with contrast agents [58, 60] (fig. 2). The tumor is usually relatively small at time of presentation. Conversely, the pineoblastoma is a high-grade tumor related to other primitive neuroectodermal tumors, such as medulloblastoma. The pineoblastoma commonly presents in children and exhibits aggressive behavior. Typically larger than pineocytomas at time of presentation, these tumors invade surrounding brain tissue and are more likely to disseminate throughout CSF pathways [14, 50]. Pineoblastomas tend to be lobulated, homogeneously enhancing with contrast agents, and are more likely to cause obstructive hydrocephalus [44] (fig. 3). The mixed pineocytoma-pineoblastoma exhibits characteristics of both tumor subtypes but behaves similarly to the more malignant pineoblastoma.

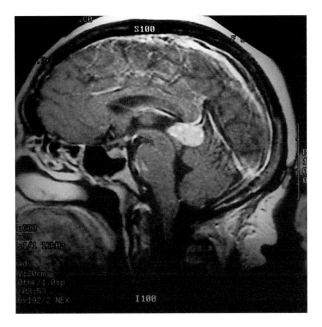

Fig. 3. Large, homogeneously enhancing mass found to be a pineoblastoma by biopsy.

Germ Cell Tumors

Germ cell tumors represent approximately 50% of all pineal region tumors. Such tumors typically develop along the midline and commonly arise in the suprasellar and pineal regions from a multipotential, as yet undetermined, precursor cell [65]. These neoplasms may be divided into germinomas and nongerminomatous germ cell tumors. Nongerminomatous tumors may be further subdivided based on their degree of differentiation. Some tumors demonstrate embryonic differentiation (embryonal carcinoma, mature/immature teratoma) while others demonstrate extraembryonic differentiation toward vitelline elements (endodermal sinus tumors) or syncytiotrophoblastic elements (choriocarcinoma) [59]. Frequently, combinations of these elements are observed and simply referred to as mixed germ cell tumors. These tumors usually present in young adults (second and third decades) with the exception of choriocarcinoma and mature teratoma, which usually present in children [55]. There is a clear gender difference with males afflicted 3–4 times more frequently than females [1, 27].

Of these tumors, germinomas are the most common, comprising approximately 61% of all germ cell tumors. Frequently, tumors with features of both germinoma and nongerminomatous tumors may be observed as may tumors with mixed nongerminomatous features. Of the nongerminomatous tumors, immature teratomas (25%) and mature teratomas (10%) are relatively common, with embryonal carcinoma (3%), endodermal sinus tumor (1.5%), and chori-ocarcinoma (1.5%) less commonly seen [56]. The tumors display a striking range of malignant behaviors and various propensities to disseminate through-out the neuroaxis via the CSF.

Gliomas

Astrocytes are normal components of both the pineal gland and the brainstem. Gliomas in this region are relatively rare and may arise directly from the pineal gland or tectal plate. Astrocytomas are more frequently en-countered than anaplastic astrocytomas and glioblastoma multiforme, but the high-grade tumors do occur [16, 20].

Meningiomas

Meningiomas in this region typically arise from the falx or tentorium. Some may arise from the tela choroidea of the third ventricle. Typically, well-circumscribed masses exhibiting slow growth, little propensity to invade surrounding brain tissue or disseminate via CSF pathways, these tumors dem-onstrate the expected homogeneous contrast enhancement on imaging which is characteristic of a meningioma [31].

The University of Pittsburgh Experience

Between April 1989 and April 1997, 14 patients with parenchymal pineal tumors or germ cell tumors of the pineal region were treated with stereotactic radiosurgery at the University of Pittsburgh. Eight patients (57%) had pineocy-tomas, 2 patients (14%) had pineoblastomas, 2 patients (14%) had germinomas, and 2 patients (14%) had nongerminomatous germ cell tumors (1 embryonal carcinoma, 1 mixed tumor). Eight patients were men (57%) and 6 were women (43%) with a mean age of 31.2 years (range 3–68) at time of treatment. All patients presented with the complaint of headache due to symptomatic hydrocephalus. Eight (57%) noted diplopia, 2 patients (14%) noted balance difficulties, and a single patient (7%) presented with confusion in addition to the headache. Of the 8 patients with diplopia, extraocular movements were abnormal on examination (including Parinaud's syndrome) in all. Twelve pa-tients (86%) underwent CSF diversion prior to radiosurgery; 8 patients (57%) required ventriculoperitoneal shunts and 4 patients (29%) underwent endo-scopic third ventriculostomy. Eleven patients (79%) had histopathologic con-firmation of their diagnosis prior to radiosurgery; 9 patients (64%) underwent stereotactic biopsy and 2 patients (14%) underwent open surgical biopsy prior to referral to our center. The remaining 3 patients (21%) were treated based on typical neurodiagnostic imaging criteria. One patient (7%) underwent subtotal resection, 2 patients (14%) had chemotherapy, and a single patient (7%) received 5,500 cGy of fractionated radiation therapy before radiosurgery.

Radiosurgical Technique

The technique of stereotactic radiosurgery using the Leksell Gamma Knife® for tumors has been previously described [47]. Briefly, the Leksell Model G stereotactic headframe was affixed to the patient's head after the administration of mild intravenous sedation and infiltration of the scalp with local anesthetic. Target localization was performed using high-resolution, contrast-enhanced, multiplanar MRI. Computer-assisted tomography was used prior to 1991. Since 1993, a specific volume acquisition MRI protocol with 1-mm slices has been used to provide detailed graphic information regarding the tumor and adjacent critical structures and to allow construction of a three-dimensional tumor volume. The images were transferred to the dose-planning computer in the radiosurgery center via a fiberoptic Ethernet system. Computerized dose planning was performed initially on a Micro-VAX II workstation (Digital Equipment Corp., Westminster, Mass., USA) and later on a Hewlett-Packard workstation using Gamma Plan® software (Elekta Instruments, Atlanta, Ga., USA). The maximum radiation dose, isodose and margin doses were all determined jointly by the neurosurgeon, radiation oncologist and medical physicist. Radiosurgery was performed using the 201 source cobalt-60 Leksell Gamma Knife® Models U or B (Elekta Instruments). Upon completion of radiosurgery, each patient received a single 40-mg intravenous dose of methylprednisolone.

Radiation Dosimetry

The mean radiosurgical tumor volume was 6.4 ml (range 1.4–14.2). Multiple isocenters were used in 13 of the patients (mean 6.1 isocenters) to conform the radiation dose to irregular tumor margins (fig. 4). All patients were treated with the 50% isodose to the tumor margin. The mean dose delivered to the tumor center was 30.7 Gy (range 24–40) and to the tumor margin was 15.4 Gy (range, 12–20). Dose selection was modified by the risk of radiation-induced complications predicted by the integrated logistic formula and history of previous radiation exposure [47].

Follow-Up

After radiosurgery, patients underwent clinical examination and serial neuroimaging with a frequency based on the individual patient's degree of

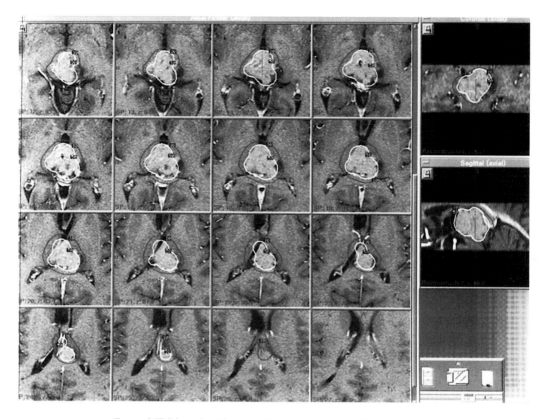

Fig. 4. MRI-based radiosurgical dose planning with precise conformation of isodose to the irregular tumor margins.

illness. In patients residing a significant distance from Pittsburgh, evaluation and imaging were carried out by the referring physician and submitted to our facility for review. Additional clinical follow-up was obtained via telephone interview.

Results

Clinical Outcome

Clinical examinations after radiosurgery were performed by the referring doctor or treating neurosurgeon in all cases. The mean length of follow-up was 29 months (range 4–99). Three patients died during this period. One patient (61 years old, pineocytoma) expired after a presumed myocardial

Table 1. Clinical outcomes of patients after radiosurgery based on tumor subtype

	Total	Better	Stable	Worse
Pineocytoma	8	2	6	0
Pineoblastoma	2	2	0	0
Germinoma	2	1	1	0
Nongerminomatous germ cell tumor				
Embryonal	1	0	0	1
Mixed germ cell	1	0	1	0

infarction at 26 months from treatment. Prior to his death, his examination had remained stable. A second patient (3 years old, pineoblastoma) died as a result of an unrelated cerebellar hemorrhage, but had shown improvement in her Parinaud's syndrome prior to death. The third patient (9 years old, embryonal carcinoma) died due to disease progression despite receiving chemotherapy and craniospinal irradiation prior to radiosurgery. Overall, the clinical neurologic examination improved in 5 patients (36%), remained stable in 8 patients (57%) and deteriorated in the 1 patient (7%) with documented tumor progression (table 1). Clinical improvement was manifest as improvement in Parinaud's syndrome and gaze paresis, or subjective improvement in diplopia. Resolution of headache, which may result from CSF diversion, was not considered neurologic improvement. One patient (35 years old, pineocytoma) did develop a new Parinaud's syndrome at 7 months from radiosurgery which improved with initiation of corticosteroids.

Neuroimaging Response

All patients underwent serial MRI scanning with a mean duration of imaging follow-up of 21 months (range 3–51). All imaging studies were reviewed at the University of Pittsburgh. Imaging demonstrated a decrease in the tumor size in 10 patients (71%), no change in tumor size in 3 patients (21%) and tumor growth in 1 patient (7%). Of the 8 tumors demonstrating a response to radiosurgery, 3 of the tumors actually resolved completely. The radiographic response may be best understood by table 2. Of the 8 patients with pineocytoma treated, 2 remained stable, 3 were smaller, and 3 disappeared completely (fig. 5, 6). Of the 2 patients with pineoblastoma treated, 1 tumor regressed and 1 disappeared. Of the 2 patients with germinoma, both tumors were smaller on follow-up imaging. Of the 2 patients with nongerminomatous

Table 2. Neuroimaging outcomes of patients after radiosurgery based on tumor subtype

	Total	Larger	Stable	Smaller	Absent
Pineocytoma	8	0	2	3	3
Pineoblastoma	2	0	0	1	1
Germinoma	2	0	0	2	0
Nongerminomatous germ cell tumor					
Embryonal	1	1	0	0	0
Mixed germ cell	1	0	1	0	0

germ cell tumors, the mixed tumor decreased in size and the embryonal carcinoma progressed locally and disseminated. MR images of the patient, who developed the delayed Parinaud's syndrome at 7 months from treatment, demonstrated subtle T2 signal changes in the brainstem and thalamus adjacent to the radiosurgical target (fig. 7).

Discussion

Given the initial morbidity associated with open biopsy and resection, the use of empiric radiation therapy seemed logical. Now, however, with the accuracy and safety of current open and stereotactic techniques, there is no longer an indication for treatment without a tissue diagnosis. The combination of tumor markers and CSF cytology may be sufficient in certain cases, however biopsy is generally recommended. Kersh et al. [40] demonstrated that histologic confirmation of germ cell tumor subtype, prior to initiation of fractionated radiation therapy, increased projected 5-year survival from 53% in unbiopsied cases to 79% in biopsied cases.

Although there is a general consensus regarding the need for tissue diagnosis, there is some debate regarding the best approach to the biopsy. Both open and stereotactic methods are effective, but the relative safety of the procedures depends upon the individual operating surgeon. The traditional microsurgical approach has gained popularity due to advanced instrumentation and techniques. Proponents of open biopsy often cite the added safety of direct visualization of the tumor and vascular structures prior to the biopsy, the possibility of curative resection, and the increased diagnostic information gained from a larger biopsy specimen [6, 24, 63, 64, 66]. One criticism of stereotactic biopsy is the limited diagnostic capacity and possible sampling error associated with

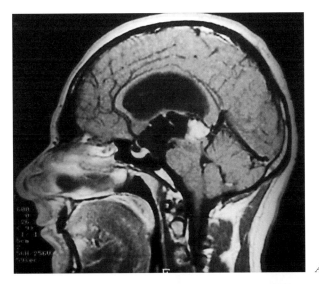

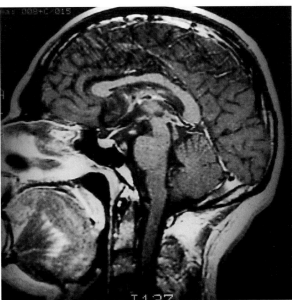

Fig. 5. Thirty-three-year-old man with pineocytoma. MRI scans from the time of radiosurgery (*A*) and 28 months postoperatively (*B*) showing shrinkage of the pineal region tumor and improvement of the hydrocephalus.

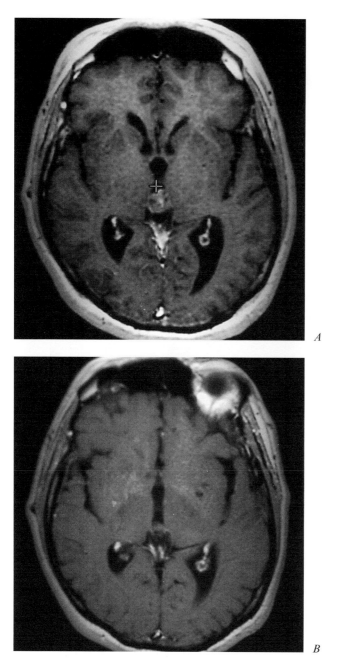

Fig. 6. Fifty-three-year-old man with pineocytoma. Axial MRI scans from the time of radiosurgery (*A*) and 24 months postoperatively (*B*) showing shrinkage of the tumor extending into the third ventricle.

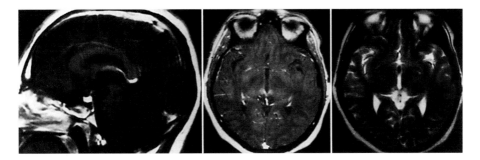

Fig. 7. Thirty-five-year-old man with pineocytoma who developed a Parinaud's syndrome 7 months after radiosurgery. MR images show shrinkage of the tumor with T2 signal change in the brainstem and thalamus adjacent to the previous target.

the relatively small specimen. Edwards and co-workers [6, 27] note that 15% of pineal region tumors contain mixed elements that may be overlooked or inadequately sampled. Also cited as shortcomings of the stereotactic technique are the risk of hemorrhage from the tumor or surrounding vascular structures and the need for a second surgical procedure (resection). Although a rare occurrence, tumor seeding due to stereotactic biopsy of a pineoblastoma has been reported [57].

Stereotactic techniques combined with advanced neuroimaging have provided a safe and effective means of obtaining tissue from deep-seated and previously inaccessible locations within the brain. The precision of the instrumentation and minimal invasiveness of the technique have made stereotactic biopsy particularly appealing. Using MRI guidance, both target and trajectory may be evaluated in three dimensions. In general, specimen size provides an accurate diagnosis, patient recovery is uncomplicated and rapid, and CSF sampling may be performed at the same time [15, 22, 42, 52, 56]. In contrast to proponents of microsurgery (who regard stereotactic biopsy as an unnecessary second procedure), supporters of stereotactic biopsy consider it a means of avoiding the risks of craniotomy in patients with tumors such as germinoma, glioma and metastatic malignancies for which a surgical benefit has not been established. Both groups will generally agree that large tumors, tumors with elevated AFP levels and tumors with evidence of ependymal or leptomeningeal spread are best managed with stereotactic biopsy.

Pineal parenchymal neoplasms have typically been classified among the malignant pineal region tumors due to their unpredictable behavior. In the treatment of pineocytomas, both surgery and radiation therapy have been advocated. Disclafani et al. [24] reported 6 cases of pineocytoma treated with 4,500 to 5,400 cGy focused radiation after subtotal resection. Two patients developed local tumor progression, while 4 patients showed long-term tumor control. Schild et al. [60] reviewed the Mayo Clinic series of 30 patients who underwent surgical resection of pineal parenchymal tumors. Four patients had gross total resections performed and 22 underwent postoperative fractionated radiation therapy. All 12 patients who received > 5,000 cGy had tumor control, while only 1 of 7 patients receiving lower radiation doses had tumor control. Additionally, there is limited evidence to suggest that resection alone may be beneficial. Vaquero et al. [67] reported long-term tumor control in patients with pineocytoma, who underwent gross total resection without radiation.

Given the more aggressive behavior of the pineoblastoma, treatment strategies have been less successful. With the use of subtotal resection followed by fractionated radiation therapy, the rate of tumor progression in one series was still 44% [60]. Two studies, also reporting the results of resection with radiation, found the postoperative median survival to be just 24 months in children and 30 months in adults [12, 50]. Chang et al. [12] also identified the presence of leptomeningeal spread to be a negative prognostic indicator that heralded a poor response to craniospinal irradiation. Because of the poor outcome, high recurrence rate and inability to treat leptomeningeal spread, multiple chemotherapy trials were undertaken. Although there are isolated case reports of favorable responses to chemotherapy, the larger pineoblastoma series have failed to substantiate a benefit. Ghim et al. [28] reported the use of a four-cycle regimen of etoposide/cisplatin/vincristine prior to craniospinal irradiation which resulted in tumor control in 2 of 3 patients. Duffner et al. [26] reported the outcome of the Pediatric Oncology Group infant brain tumor study, which treated 11 patients with pineoblastoma. All patients received 2 cycles of cyclophosphamide/vincristine followed by 1 cycle of cisplatin/etoposide and craniospinal irradiation. Following chemotherapy, only 1 patient demonstrated a partial response, while 5 remained stable and 5 exhibited tumor progression. All patients went on to fail treatment with survival ranging from 4 to 13 months. The chemotherapy regimen was neither effective in controlling local tumor growth, nor in preventing leptomeningeal spread of disease. Jakacki et al. [36] reported the outcome of the Children's Cancer Group study. Twenty-five children with pineoblastoma were treated with craniospinal radiation and/or chemotherapy. Eight infants under 18 months were treated with 8 chemotherapeutic agents in 1 day (8-in-1 regimen) without radiation. The remaining 17 children were treated with radiation then randomized to either vincristine/lomustine/prednisone or the 8-in-1 regimen. Despite treatment, all infants developed disease progression at a median of 4 months from diagnosis. The older children fared slightly better with a 61% progression-free survival at 3 years, representing a significant improvement over the natural history of the disease.

There are multiple reports regarding the radiosensitivity of germinoma, with tumor control rates ranging from 72 to 89% and 5-year survival rates of 72–100% [1, 2, 18, 19, 33, 35, 37]. The average treatment dose of 5,000 cGy has been delivered as both a local dose and as whole-brain irradiation with a coned boost to the tumor bed. Most treatment failures occur within 2 years of initiation of therapy and may represent local recurrence, spinal leptomeningeal spread or unrecognized mixed tumor elements. Most groups recommend whole-brain radiation of approximately 2,500 cGy followed by a coned boost to the pineal region of an additional 2,500 cGy. Spinal irradiation has generally been reserved for those patients with positive CSF cytology or neuroimaging evidence of tumor seeding.

The desire to limit the necessary radiation dose in these patients has led to the use of induction chemotherapy. Previously reserved for nongerminomatous germ cell tumors, chemotherapy appears to be effective in conjunction with a decreased radiation dose. Sebag-Montefiore et al. [61], utilizing a regimen of vincristine/etoposide/carboplatin to treat 7 patients with germinoma, demonstrated a decrease in tumor size in 6 patients prior to radiation. All patients were disease-free at a median 12 months from treatment. Allen et al. [3] treated 11 patients with germ cell neoplasms (7 germinoma, 4 nongerminomatous germ cell tumors). All patients received neoadjuvant chemotherapy with either cyclophosphamide alone (germinoma) or a combination of cyclophosphamide/vinblastine/bleomycin/cisplatin (nongerminoma). Ten of the 11 patients had a complete response while 1 had a partial response to chemotherapy prior to reduced dose irradiation. Ten patients remained disease-free at a median of 47 months from treatment.

The Role of Radiosurgery

The safety and efficacy of radiosurgery has been established for a wide spectrum of intracranial neoplastic and vascular lesions. However, there has been relatively little reported on the use of radiosurgery for tumors of the pineal region. Backlund et al. [5], in 1974, documented the use of radiosurgery for 2 patients with pineocytomas. On imaging follow-up, both patients had complete resolution of the tumor. In 1990, Casentini et al. [11] described radiosurgery for the treatment of 4 patients with germinoma. These tumors also showed an excellent response to radiosurgery without significant complications. In 1992, Dempsey and Lunsford [23] reported the early University of Pittsburgh experience with radiosurgery for pineal region tumors, most of which were gliomas and meningiomas. Of the 9 patients treated, 6 demonstrated a decrease in tumor size, while the tumors in 3 patients remained stable over the period of follow-up. The current series represents the largest series of patients with pineal parenchymal and germ cell tumors of the pineal region treated with radiosurgery to date. The tumor control rate is equal or superior to that reported in other series with combinations of resection, fractionated radiation therapy and chemotherapy. Radiosurgery is particularly well suited to lesions of the pineal region. Current imaging allows precise definition of tumor location and volume as well as radiosurgical dose planning. Multiple isocenters are used to construct a dose plan which conforms well to the tumor margin but avoids adjacent critical neurovascular structures.

One of the keys to the effectiveness of radiosurgery remains appropriate patient selection. The only alternatives previously, surgery and fractionated radiation therapy, exhibited significantly higher risks to the patient. Bruce and Stein [9] report the operative mortality for surgery of the pineal region to reach 8% with the risk of significant morbidity as high as 12%. In the series from the New York Neurological Institute, 169 operations for pineal region tumors were performed. Patients with benign tumors did well (58 of 59 patients had excellent outcomes) with a 90% rate of gross total resection. Those with malignant tumors fared poorly with a 15% risk of major morbidity and only a 30% chance of complete resection [8]. Kanno [39] operated on 30 patients with pineocytoma via an infratentorial approach with an excellent outcome in 26 (87%) with a mortality rate of 10%. Nazzaro et al. [53] reported the results from the transtentorial approach to 12 pineal region tumors. Although only 4 patients presented with gaze problems related to midbrain dysfunction, all patients had evidence of Parinaud's syndrome and other cranial nerve deficits after operation. Similarly, all patients presented with full visual fields, yet all 12 patients had a complete or partial homonymous hemianopsia post-operatively.

The risks of fractionated radiation, particularly upon the developing brain, have been well documented. Radiation-induced complications including visual loss, parenchymal necrosis and deterioration of intellectual function have been reported at rates of 10–38% [18, 29, 48, 51]. Obviously, in patients with large tumors (>30 mm diameter) or tumors with evidence of local invasion or CSF dissemination, biopsy and fractionated radiation therapy are better options. Conversely, in patients having benign tumors, without contraindications to surgery, resection remains the treatment of choice. However, in patients with tumors of appropriate size and favorable histology, or patients with contraindications to open resection, stereotactic radiosurgery remains an effective treatment alternative. To better define a strategy for approaching these tumors, it is possible to construct a treatment algorithm (fig. 8). Clearly, treatment must be adapted to the individual patient, however, such an algorithm may represent a starting point from which to build.

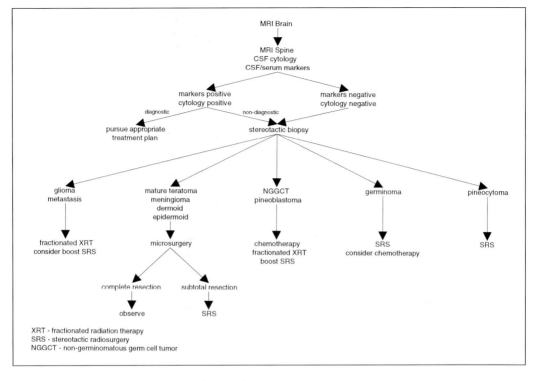

Fig. 8. Treatment algorithm for pineal region tumors.

Conclusion

Management of pineal region tumors remains a significant challenge in the field of neurosurgery. By using the appropriate combination of available treatment modalities, we may be able to improve the quality of life and length of survival of more patients. Although constrained by the usual arguments of limited patient population and length of follow-up, we believe stereotactic biopsy and radiosurgery to be two options which represent safe and effective alternatives to the conventional approaches of resection and fractionated radiation therapy.

References

1　Abay EO, Laws ER, Grado GL: Pineal tumors in children and adolescents. J Neurosurg 1981;55: 889–895.

2　Allen J: Management of primary intracranial germ cell tumors of childhood. Pediatr Neurosci 1987; 13:152.

3　Allen J, Kim J, Packer R: Neoadjuvant chemotherapy for newly diagnosed germ cell tumors of the central nervous system. J Neurosurg 1987;67:65.

4　Ashley DM, Longee D, Tien R: Treatment of patients with pineoblastoma with high dose cyclophosphamide. Med Pediatr Oncol 1996;26:387–392.

5　Backlund EO, Rahn T, Sarby B: Treatment of pinealomas by stereotaxic radiation surgery. Acta Radiol (Ther) 1974;13:368–376.

6　Baumgartner JE, Edwards MS: Pineal tumors. Neurosurg Clin North Am 1992;3:853–862.

7　Brada M, Laing R: Radiosurgery/stereotactic external beam radiotherapy for malignant brain tumours: The Royal Marsden Hospital experience. Recent Res Cancer Res 1994;135:91–104.

8　Bruce JN, Stein BM: Pineal tumors. Neurosurg Clin North Am 1990;1:123–138.

9　Bruce JN, Stein BM: Surgical management of pineal region tumors. Acta Neurochir 1995;134: 130–135.

10　Camins MB, Schlesinger EB: Treatment of tumours of the posterior part of the third ventricle and the pineal region: A long-term follow-up. Acta Neurochir (Wien) 1978;40:131–143.

11　Casentini L, Colombo F, Pozza F: Combined radiosurgery and external radiotherapy of intracranial germinomas. Surg Neurol 1990;34:79–86.

12　Chang SM, Lillis-Hearne PK, Larson DA, Wara WM, Bollen AW, Prados MD: Pineoblastoma in adults. Neurosurgery 1995;37:383–390.

13　Chapman PH, Linggood RM: The management of pineal area tumors: A recent reappraisal. Cancer 1980;46:1253–1257.

14　Chiechi MV, Smirnitopoulos JG, Mena H: Pineal parenchymal tumors: CT and MR features. J Comput Assist Tomogr 1995;19:509–517.

15　Conway LW: Stereotaxic diagnosis and treatment of intracranial tumors including an initial experience with cryosurgery for pinealomas. J Neurosurg 1973;38:453–460.

16　Cummins FM, Taveras JM, Schlesinger EB: Treatment of gliomas of the third ventricle and pinealomas. Neurology 1960;10:1031–1036.

17　Dandy WE: Operative experience in cases of pineal tumor. Arch Surg 1936;33:19–46.

18　Dattoli MJ, Newall J: Radiation therapy for intracranial germinoma: The case for limited volume treatment. Int J Radiat Oncol Biol Phys 1990;19:429–433.

19　Dearnaley D, A'Hern R, Whittaker S, Bloom HJ: Pineal and CNS germ cell tumors: Royal Marsden Hospital experience 1962–1987. Int J Radiat Oncol Biol Phys 1990;18:773–781.

20　DeGirolami U, Schmidek H: Clinicopathological study of 53 tumors of the pineal region. J Neurosurg 1973;39:455–462.

21　Demakas JJ, Sonntag VKH, Kaplan AM: Surgical management of pineal area tumors in early childhood. Surg Neurol 1982;17:435–440.

22　Dempsey PK, Kondziolka D, Lunsford LD: Stereotactic diagnosis and treatment of pineal region tumors and vascular malformations. Acta Neurochir 1992;116:14–22.

23　Dempsey PK, Lunsford LD: Stereotactic radiosurgery for pineal region tumors. Neurosurg Clin North Am 1992;3:245–253.

24　Disclafani A, Hudgins RJ, Edwards MS, Wara W, Wilson CB, Levin VA: Pineocytomas. Cancer 1989;63:302–304.

25　Donat JF, Okazaki H, Gomez MR: Pineal tumors: A 53-year experience. Arch Neurol 1978;35: 736–740.

26　Duffner PK, Cohen ME, Sanford RA, Horowitz ME, Krischer JP, Burger PC, Friedman HS, Kun LE: Lack of efficiency of postoperative chemotherapy and delayed radiation in very young children with pineoblastoma. Pediatric Oncology Group. Med Pediatr Oncol 1995;25:38–44.

27　Edwards MSB, Hudgins RJ, Wilson CB: Pineal region tumors in children. J Neurosurg 1993;68: 689–697.

28   Ghim TT, Davis P, Seo JJ, Crocker I, O'Brien M, Krawiecki N: Response to neoadjuvant chemother-
      apy in children with pineoblastoma. Cancer 1993;72:1795–1800.
29   Gutin P, Leibel S, Sheline G: Radiation Injury to the Nervous System. New York, Raven Press,
      1991.
30   Herrmann HD, Westphal M, Winkler K, Laas RW, Schulte FJ: Treatment of nongerminomatous
      germ cell tumors of the pineal region. Neurosurgery 1994;34:524–529.
31   Herrmann HD, Winkler D, Westphal M: Treatment of tumors of the pineal region and posterior
      part of the third ventricle. Acta Neurochir 1992;116:137–146.
32   Hitchon PW, Abu-Yousef MM, Graf CJ: Management and outcome of pineal region tumors.
      Neurosurgery 1983;13:248–253.
33   Horowitz M, Hall W: Central nervous system germinomas: A review. Arch Neurol 1991;48:652.
34   Horrax G: Treatment of tumors of the pineal body. Arch Neurol Psychiatry 1950;64:227–242.
35   Huh SJ, Shin KH, Kim IH, Ahn YC, Ha SW, Park CI: Radiotherapy of intracranial germinomas.
      Radiother Oncol 1996;38:19–23.
36   Jakacki RI, Zeltzer PM, Boyett JM, Albright AL, Allen JC, Geyer JR, Rorke LB, Stanley P, Stevens
      KR, Wisoff J: Survival and prognostic factors following radiation and/or chemotherapy for primitive
      neuroectodermal tumors of the pineal region in infants and children: A report of the Children's
      Cancer Group. J Clin Oncol 1995;13:1377–1383.
37   Jenkins RD, Simpson WJK, Keen CW: Pineal and suprasellar germinomas: Results of radiation
      treatment. J Neurosurg 1978;48:99–107.
38   Jooma R, Kendall BE: Diagnosis and management of pineal tumors. J Neurosurg 1983;58:654–665.
39   Kanno T: Surgical pitfalls in pinealoma surgery. Minim Invasive Neurosurg 1995;38:153–157.
40   Kersh CR, Constable WC, Eisert DR, Spaulding CA, Hahn SS, Jenrette JM, Marks RD: Primary
      central nervous system germ cell tumors. Effect of histologic confirmation on radiotherapy. Cancer
      1988;61:2148–2152.
41   Kobayashi T: Combination chemotherapy with cisplatin and etoposide for malignant intracranial
      germ cell tumors. J Neurosurg 1989;70:676.
42   Kreth FW, Schatz CR, Pagenstecher A, Faist M, Volk B, Ostertag CB: Stereotactic management
      of lesions of the pineal region. Neurosurgery 1996;39:280–289.
43   Kretschmar CS: Germ cell tumors of the brain in children: A review of current literature and new
      advances in therapy. Cancer Invest 1997;15:187–198.
44   Lambrinides K, Reichert M: MR imaging of pineoblastomas. Radiol Technol 1994;66:106–110.
45   Lapras C, Parer JD, Mottolese C, Lapras C Jr: Direct surgery for pineal tumors: Occipital-transten-
      torial approach. Prog Exp Tumor Res 1987;30:268–280.
46   Lingood R, Chapman P: Pineal tumors. J Neurooncol 1992;12:85.
47   Lunsford LD, Flickinger J, Coffey RJ: Stereotactic gamma knife radiosurgery. Initial North Amer-
      ican Experience in 207 patients. Arch Neurol 1990;47:169–175.
48   Marsh WR, Laws ER: Shunting and irradiation of pineal tumors. Clin Neurosurg 1985;32:384–396.
49   Matsutani M, Sano K, Takakura K, Fujimaki T, Nakamura O, Funata N, Seto T: Primary intracranial
      germ cell tumors: A clinical analysis of 153 histologically verified cases. J Neurosurg 1997;86:446–455.
50   Mena H, Rushing EJ, Ribas JL, Delahunt B, McCarthy WF: Tumors of pineal parenchymal cells:
      A correlation of histological features, including nucleolar organizing regions, with survival in 35
      cases. Hum Pathol 1995;26:20–30.
51   Moris JG, Grattan-Smith P, Panegyres PK, O'Neill P, Soo YS, Langlands AO: Delayed cerebral
      radiation necrosis. Q J Med 1994;87:199–229.
52   Moser RP, Backlund EO: Stereotactic techniques in the diagnosis and treatment of pineal region
      tumors; in Neuwelt EA (ed): Diagnosis and Treatment of Pineal Region Tumors. Baltimore, Williams
      & Wilkins, 1984, pp 236–253.
53   Nazzaro JM, Shults WT, Neuwelt EA: Neuro-ophthalmological function of patients with pineal
      region tumors approached transtentorially in the semisitting position. J Neurosurg 1992;76:746–751.
54   Neuwelt EA: An update on the surgical treatment of malignant pineal region tumors. Clin Neurosurg
      1985;32:397–428.
55   Neuwelt EA (ed): Diagnosis and Treatment of Pineal Region Tumors. Baltimore, Williams & Wilkins,
      1984.

56 Pecker J, Scarabin JM, Vallee B: Treatment of tumors of the pineal region: Value of stereotaxic biopsy. Surg Neurol 1979;12:341–348.
57 Rosenfeld J, Murphy M, Chow C: Implantation metastasis of pineoblastoma after stereotactic biopsy: Case report. J Neurosurg 1990;73:287.
58 Satoh H, Uozumi T, Kiya K, Kurisu K, Arita K, Sumida M, Ikawa F: MRI of pineal region tumors: Relationships between tumours and adjacent structures. Neuroradiology 1995;37:624–630.
59 Scheithauer B: Neuropathology of pineal region tumors. Clin Neurosurg 1984;32:351.
60 Schild SE, Scheithauer BW, Schomberg PJ, Hook CC, Kelly PJ, Frick L, Robinow JS, Buskirk SJ: Pineal parenchymal tumors. Clinical, pathological, and therapeutic aspects. Cancer 1993;72:870–880.
61 Sebag-Montefiore DJ, Douek E, Kingston JE, Plowman PN: Intracranial germ cell tumours: Experience with platinum-based chemotherapy and implications for curative chemoradiotherapy. Clin Oncol 1992;4:345–350.
62 Shibamoto Y, Oda Y, Yamashita J, Takahashi M, Kikucki H, Abe M: The role of cerebrospinal fluid cytology in radiotherapy planning for intracranial germinoma. Int J Radiat Oncol Biol Phys 1994;29:1089–1094.
63 Stein BM, Bruce JN: Surgical management of pineal region tumors. Clin Neurosurg 1992;39: 509–532.
64 Stein BM, Fetell MR: Therapeutic modalities for pineal region tumors. Clin Neurosurg 1985;32: 445–455.
65 Sumida M, Uozumi T, Kiya K, Mukada K, Arita K, Kurisu K, Sugiyama K, Onda J, Satoh H, Ikawa F: MRI of intracranial germ cell tumors. Neuroradiology 1995;37:32–37.
66 Van Wagenen WP: A surgical approach for the removal of certain pineal tumors: Report of a case. Surg Gynecol Obstet 1931;53:216.
67 Vaquero J, Ramiro J, Martinez R, Coca S, Bravo G: Clinicopathological experience with pineocytoma: Report of five surgically treated cases. Neurosurgery 1990;27:612.
68 Wara W, Jenkin D, Evans A: Tumors of the pineal and suprasellar region: Children's Cancer Study Group Treatment Results 1960–1975. Cancer 1979;43:698.

Brian R. Subach, MD, University of Pittsburgh Medical Center, Suite B–400,
Department of Neurological Surgery, 200 Lothrop Street, Pittsburgh, PA 15213 (USA)
Tel. (412) 647 3685, Fax (412) 647 0989

Lunsford LD, Kondziolka D, Flickinger JC (eds): Gamma Knife Brain Surgery.
Prog Neurol Surg. Basel, Karger, 1998, vol 14, pp 195–211

....

Gamma Knife Radiosurgery for the Treatment of Movement Disorders

Christopher M. Duma[a], *Deane B. Jacques*[a], *Oleg Kopyov*[a],
Rufus J. Mark[b], *Brian Copcutt*[a], *Melissa Gembus*[a], *Halle K. Farokhi*[a]

[a] The Neurosciences Institute, and [b] Department of Radiation Oncology,
Good Samaritan Hospital, Los Angeles, Calif., USA

The original purpose of the 'Gamma knife', designed and created by Prof. Lars Leksell [31], was to be able to treat functional disorders by making a small, well-circumscribed lesion within the brain without opening the skull and without the need for heat or mechanical destructive forces. Functional disorders have been treated using stereotactic radiosurgery as far back as 1951, when Prof. Leksell [31] experimentally treated trigeminal neuralgia with an orthovoltage x-ray tube attached to his stereotactic arc-centered system

The prototype cobalt-60 Gamma unit was developed in 1967, enabling precise and accurate targeting with very small collimators. Because of its extraordinary mechanical accuracy, the Gamma unit was ideally suited for the field of stereotactic functional radiosurgery. The units' 4-mm collimator, which at its target made a 5- to 6-mm lesion, lent itself to the goal of noninvasive destruction of aberrant neural pathways. The progress of radiosurgery, however, was impeded by the difficulties related to precise targeting for functional disorders without physiological feedback.

By the 1980s, radiofrequency thermocoagulation of various targets within the ventrolateral thalamus had become the preferred method of surgical amelioration of movement disorders related to Parkinson's disease (PD) [26, 39]. Meanwhile, the prototype Gamma unit and its users had been accumulating years of experience treating arteriovenous malformations and trigeminal neuralgia, and the surgical community was gaining a better understanding and increased knowledge of the radiobiology of single-fraction radiosurgery. The indications for using a tool that had no means for intraoperative physiological feedback, however, had to be strict and thus experience with functional radiosurgery began only slowly [7].

As the fields of functional neurosurgery and radiosurgery grew, however, so did the *non*-surgical treatment of many of the movement disorders [2, 29, 32, 43]. More effective medications for pain, PD, and psychiatric illnesses, all but halted the progress of stereotactic functional neurosurgery. These factors compounded with the lack of physiological feedback inherent in the technique caused an interest in the use of radiosurgery for the precise targeting of functional disorders to wane. Therefore, information regarding radiosurgical lesion size and predictability within the deep structures of the brain is limited.

With recent improvements in neuroimaging, and the development of third- and fourth-generation radiosurgical dose-planning software, came a renewed interest in using radiosurgery for the treatment of movement disorders, specifically in PD patients who have conditions which predispose them to the risks of *invasive* stereotactic neurosurgery. This subgroup of patients include those who are taking anticoagulants, have respiratory or cardiac disease, are very elderly or are generally poor risks for surgery. In addition, many patients, when offered the choice, prefer a less invasive alternative.

Radiosurgery involves no opening of the cranium and no incisions, eliminating both the risk of hemorrhage from passing an electrode to the depths of the thalamus and pallidum as well as the potential risk of meningitis from operative infection. It is for these reasons stereotactic radiosurgical treatment of movement disorders has value in a small subgroup of patients.

Contained in this chapter is a 6-year retrospective review of Gamma knife radiosurgical thalamotomy in 34 patients performed at Good Samaritan Hospital in Los Angeles, California. This is the largest reported series of patients with the longest median follow-up to date. A comparison of 'low-dose' radiosurgical lesions versus 'high-dose' radiosurgical lesions is also detailed. Included in this review are 7 patients described by us in a previous communication [41].

Also, early in our experience we created lesions in the globus pallidus interna (GPi) for treatment of the rigidity and bradykinesia of PD. Contrary to our excellent functional results and minimal complications of Gamma knife nucleus ventralis intermedius (VIM) thalamotomy for the tremor of PD [8], we had discouraging results with Gamma knife *pallidotomy*. Our retrospective experience of Gamma pallidotomy in 18 patients is also reported in this chapter. It will be evident that the reproducibility, precision and accuracy of radiosurgical pallidotomy differs markedly from that of radiosurgical thalamotomy.

Table 1. Clinical conditions predisposing a patient to radiosurgery

Clinical condition	Percent of patients
Patient choice	53
Taking anticoagulants	15
Advanced age	15
Severe cardiopulmonary disease	12
Mild dementia	5

Materials and Methods

Between March 1991 and December 1996, 34 patients (24 males and 10 females) with disabling tremor from PD recalcitrant to medical therapy underwent stereotactic radiosurgery using the 201-source cobalt-60 Gamma knife at Good Samaritan Hospital. Four patients had bilateral procedures separated by 6 months for a total of 38 lesions. Twenty-nine lesions were left-sided for right-sided tremor. No patients had prior surgery for their PD symptoms. For purposes of dose-response comparison, two treatment subgroups within the thalamotomy group were retrospectively defined. Sixteen patients (43%) were considered to have received 'high-dose' prescriptions (maximum target dose 160 Gy mean; range 140–160). Twenty-two patients (57%) were defined as the 'low-dose' group (120 Gy mean; range 110–135).

Between March 1991 and August 1995, 18 patients with medically recalcitrant and disabling bradykinesia and rigidity of PD underwent stereotactic radiosurgical pallidotomy. Patient ages ranged from 59 to 85 at time of treatment, median 73 years old. No patients had had prior surgery for their PD symptoms. All pallidotomy patients were tested both off and on levodopa (challenge test), and were accepted for treatment only if responsive to levodopa.

Inclusion criteria for radiosurgery also dictated that patients were poor surgical or anesthetic risks, had advanced age, used anticoagulants, or made a personal choice to avoid invasive surgery (table 1). The United Parkinson's Disease Rating Scale (UPDRS) [9] was used to assess the patients pre- and postoperatively. Only patients with UPDRS grade 3 or 4 disability were chosen for treatment.

Target localization of VIM thalamus was determined by coordinates based on the position of the nucleus relative to the AC/PC line, anatomical information gathered from very-high-resolution magnetic resonance imaging (MRI) and subjective surgeon correlation with the Schaltenbrand atlas [42]. The 50% isodose line of the 4-mm collimator was placed at the edge of the contralateral internal capsule medially (fig. 1). The average target coordinates for all thalamotomies in this series was $X = 15$ mm lateral to the AC/PC line, $Y = 6$ mm posterior to the midpoint of the AC/PC line, and $Z = 4$ mm superior to the AC/PC line. Target localization of the GPi was determined also by coordinates based on anatomical information gathered from high-resolution MRI and subjective surgeon correlation with the Schaltenbrand atlas [42]. The 50% isodose line of a single or double isocenter 4-mm collimator plan was placed at the center of GPi (fig. 2). Fifteen patients were treated using a single

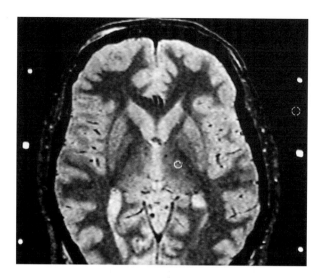

Fig. 1. Stereotactic VIM target planning MRI, STIR-weighted sequence (Gamma Plan, Elekta Instruments) in the axial plane.

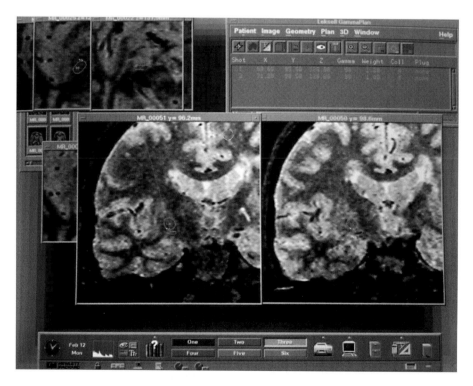

Fig. 2. Radiosurgically planned target for GPi.

Table 2. Correlation of independent neurologist evaluations and patient self-assessment scores

ΔUPDRS score as reported by independent neurologists	Subjective improvement as reported by the patients
0	None
1	Mild (1–33% relief)
2	Good (34–66%)
3	Excellent (67–99%)
3 or 4	Absent tremor

4-mm collimator with a median maximum prescription dose of 160 Gy (range 90–165). Three patients were treated using a combination of two 4-mm shots with a dose of 160 Gy.

Independent neurologist evaluations and UPDRS scoring of patient response to treatment were obtained at regular clinical follow-up intervals particular to the referring neurologists. Patients underwent 1-, 3-, and 6-month follow-up examinations by the treating neurosurgeons, but these evaluations were not used in the outcome data. Patients were also asked to subjectively rate the percentage improvement of their symptoms on a 'UPDRS-correlated' improvement scale. Statistical correlation between patient and independent neurologist assessment of outcome was made using the Pearson correlation analysis. Patient clinical outcome per low-dose versus high-dose groups was statistically analyzed using the Wilcoxon nonparametric test.

'Mild' improvement was categorized as a change of one UPDRS grade per independent neurologist evaluation and a subjective patient response of '1–33% improved'. 'Good' improvement was categorized as a change of two UPDRS grades and 34–66% subjective patient response improvement, and 'excellent' improvement was categorized as a change of three UPDRS grades and a subjective patient response score of 67–99% improvement (table 2).

The UPDRS tremor score was the only objective scoring parameter followed in all of the 34 thalamotomy patients. In 7 of these patients, tests of general PD functions were also assessed: 'Observed activity', 'activity by history', 'time to walk 20 feet' and '20-second time tests'. These were analyzed using the paired two-tailed Student's t-test. Subjective assessment of visual status and cognitive function were obtained from the patient and the examining neurologist.

The Leksell Model 'G' stereotactic coordinate frame (Elekta Instruments, Atlanta, Ga., USA) was attached to the patient's head prior to high-resolution MRI using 2-mm thick slices in axial and coronal plains. T1-, T2- and STIR-weighted images were used for target localization. Single and double isocenter computer dose planning was performed using the Gamma Plan© software system (Elekta Instruments). The maximum treatment dose was determined jointly by the attending neurosurgeon, radiation oncologist and medical physicist. The selection of dose was guided by a dose-volume analysis of prior experience with parenchymal tolerance to Gamma knife radiosurgery. In selected cases one or more of the 201 collimator sources were blocked to reduce exposure to the lens.

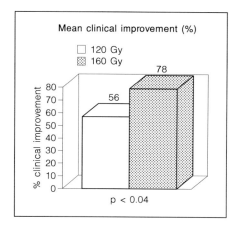

Fig. 3. Percentage improvement in tremor between high- and low-dose treatment groups. The high-dose group (160 Gy mean, range 140–165) had significantly better tremor control than the low-dose (120 Gy mean) group (p < 0.04).

Immediately after radiosurgery, all patients were given a single intravenous dose of 10 mg of dexamethasone. All patients were discharged within 36 h.

Follow-up MRI was performed at 3-month intervals for the first 6 months and then at 6-month intervals thereafter. MRI protocols included 2-mm high-resolution axial and coronal T2- and T1-weighted images with and without gadolinium. Differences in the MRI lesion size between the low-dose (120 Gy) and high-dose (160 Gy) groups within the thalamotomy group were analyzed using the unpaired t-test, ANOVA and the Wilcoxon nonparametric test.

Results

Gamma Knife Thalamotomy

Tremor Relief. Clinical and radiological follow-up ranged from 6 to 58 months (median 28). Changes in clinical tremor as determined by the Δ UPDRS scoring by the neurologists, and by the objective scoring by the patients were highly correlative: 0.89 (Pearson correlation coefficient, p < 0.001).

No change in tremor occurred in 4 Gamma knife thalamotomies (10.5%), 'mild' improvement was seen in 4 (10.5%), 'good' improvement was seen in 11 (29%), and 'excellent' improvement in 10 (26%). In 9 thalamotomies (24%), the tremor was eliminated completely. The high-dose (160 Gy mean maximum dose) thalamotomy lesion was more effective at reducing tremor (78% mean improvement) than the low-dose (120 Gy mean maximum dose) lesion (56% mean improvement, p < 0.04, Wilcoxon nonparametric test) (fig. 3).

Median time of onset of improvement was 2 months (range 1 week to 8 months). Two patients who underwent unilateral thalamotomy had bilateral improvement of their tremors. Two patients who had initial improvement in

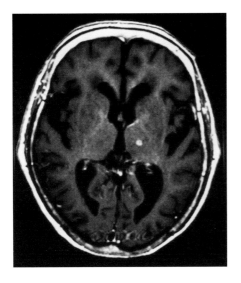

Fig. 4. 6-month follow-up T1-weighted MR image with gadolinium enhancement.

their tremors but who eventually returned to baseline in their follow-up were included in the treatment failure group. All other patients maintained their level of improvement throughout the course of the follow-up. The 4 patients who received bilateral thalamotomies separated by a 6-month interval had no subjective cognitive or performance changes other than improvement in their tremors.

Seven patients underwent formal testing of general tests of overall function. The 'observed activity' scores post-Gamma knife thalamotomy were improved ($p < 0.02$), as were the 'activity by history' score ($p < 0.05$). The 'time to walk 20 feet' scores and the '20-second timed test' scores did not statistically differ between the patients.

There were no neurological complications. No objective adverse changes in visual fields, or subjective changes in cognitive function, or performance occurred as a result of treatment.

Radiologic Findings. MRI showed a circumscribed spherical lesion which enhanced with gadolinium on T1-weighted images at a median of 3 months after radiosurgical lesioning, and a mildly diffuse T2 signal change which usually followed white matter tracts, at a median of 4.5 months following treatment, representing edema.

The average T1-weighted, gadolinium-enhancing lesion size was no different for the low- and high-dose groups and ranged from 3 to 6 mm (mean 5.0) at a median follow-up of 6 months. This lesion was present on follow-up scans as far out as 58 months (fig. 4).

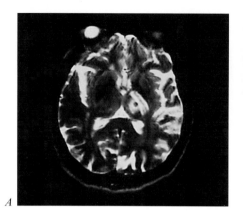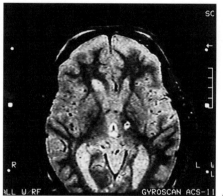

A   *B*

Fig. 5. T2-weighted sequences at 6-month follow-up showing a comparison of the extremes of edema patterns between the high dose (*A*) (160 Gy) and the low dose (*B*) (120 Gy) groups. Note the 'white matter streaking' seen usually in the high-dose group. The tremor had disappeared in the patient on the left, and there were no clinical sequelae of the T2-weighted changes.

The average T2-weighted lesion size was no different for the low- and high-dose groups and ranged from 6 to 22 mm (mean 9.2) at a median of 6 months' follow-up. This lesion also persisted on future MRIs (fig. 5). Although there was a trend toward more edema in the 160-Gy treatment group, the differences in the T1- and T2-weighted images of the thalamic lesions between the two groups did not significantly differ (fig. 6).

Gamma Knife Pallidotomy

Clinical Effects and Complications. Clinical and radiological follow-up ranged from 6 to 40 months (median 8). Only 6 patients (33%) showed transient improvement in rigidity and dyskinesia. Three patients (17%) were unchanged, and 9 patients (50%) were worsened by the treatment (table 3). Of the 6 patients with improvement, 2 exhibited visual field deficits. Overall, 4 (22%) patients had a visual field deficit, 3 patients had speech and/or swallowing difficulties, 3 had worsening of their gait, and 1 had numbness in the contralateral hemibody. Nine patients (50%) had one or more complications related to treatment which were unresponsive to steroid treatment and considered permanent.

Radiologic Findings. Lesion size on follow-up MRIs was extremely variable. For the same dose at similar follow-up intervals (160 Gy maximum dose at 8-month follow-up) lesion sizes varied from 6 to 30 mm on T1-weighted MR sequences with gadolinium enhancement (fig. 7A–C). Immeasurable variability in edema patterns was visible at the same follow-up intervals on T2-

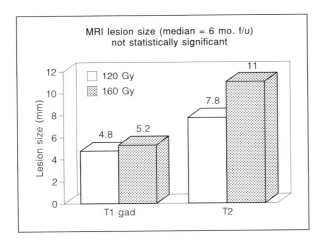

MRI lesion size (median = 6 mo. f/u)
not statistically significant

Fig. 6. Lesion sizes on T1-weighted sequences with gadolinium and on T2 weighted sequences between the low- and high-dose groups. There was no significant difference. (T1-weighted sequences: mean diameter 4.8 ± 0.6 and 5.2 ± 0.7 mm, respectively; T2-weighted sequences: mean 7.8 ± 1.3 and 11 ± 4.2 mm, respectively.)

Table 3. Gamma knife radiosurgical pallidotomy for PD

Patient	Follow-up, months	Rigidity, dyskinesia	Complication(s)
WA	8	No change	None
AB	7	No change	None
HM	35	No change	None
LL	9	Improved	None
MA	13	Improved	None
JD	29	Improved	Homonymous visual field cut
LG	40	Improved	None
SG	7	Improved	Homonymous visual field cut
MW	36	Improved	None
RG	24	Worse	Homonymous visual field cut
CK	6	Worse	None
KR	7	Worse	Dysphagia, dysarthria, hemiparesis
WS	7	Worse	Homonymous visual field cut, dysphagia
BV	4	Worse	Hemianesthesia
MP	12	Worse	None
JT	12	Worse	Dysphagia, dysarthria
MT	6	Worse	Worse gait
WW	6	Worse	Worse gait

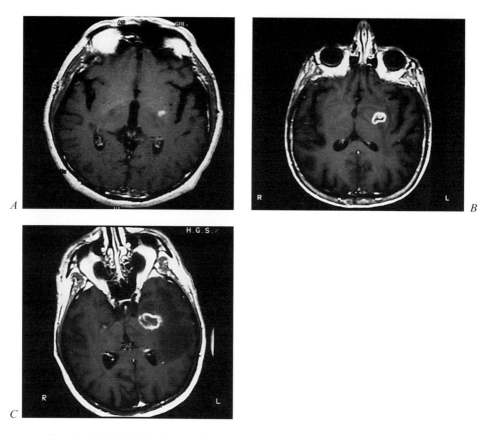

Fig. 7. A–C Variability in gadolinium-enhanced T1-weighted 8-month follow-up MR images between 3 different patients who received the same 160-Gy maximum prescription dose with a single 4-mm collimator.

weighted MR sequences (fig. 8A–C). Over time, lesion sizes tended to decrease slightly but in general were consistent throughout the course of follow-up.

Discussion

Collimator and Dose Selection

Leksell, Lindquist and co-workers [28, 29, 31–33] have described radiosurgical necrotic lesions from their early experience with functional radiosurgery. Doses of 160–200 Gy maximum were felt to be effective at creating a permanent necrotic lesion. Large collimators were to be avoided based on complications experienced in their early experience.

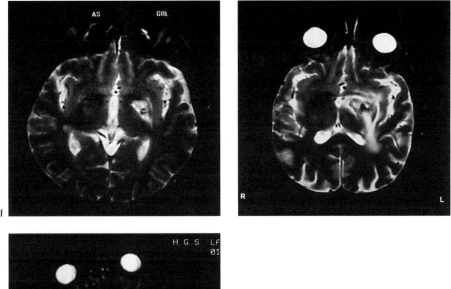

Fig. 8. A–C Extreme edema variability in T2-weighted 8-month follow-up MR images between 3 different patients who received the same 160-Gy maximum prescription dose with a single 4-mm collimator.

Because the nature of the radiosurgical lesion was still being elucidated clinically, we tended in our early experience, toward using smaller maximum doses for reasons of patient safety. Based on previous reports of the reliability of 4-mm collimator irradiation of the rat brain and its dose-response relationship for the parenchyma [20], patients treated later in our experience tended to receive higher radiosurgical doses. As time went on, more experimental information became available. Kondziolka et al. [20] reported their results in experimental rat brain lesioning and baboon lesioning [pers. commun.], and our doses were increased with the hope of improved clinical effect.

Radiosurgical Thalamotomy

Our series represents the longest follow-up, for the most patients selected for radiosurgical thalamotomy using Gamma knife technique. Overall, 34 of 38 (89%) radiosurgical thalamic lesions were effective in reducing or eliminating tremor. In 24% of patients the tremor was abolished completely, and in another 55% the patients achieved an average improvement of their UPDRS score by 2 or 3 grades. Independent neurologist evaluation correlated highly with the subjective reports of the patients themselves. No complications occurred.

Our results are similar to those previously reported. In the late 1980s the first 2 patients were treated for the tremor of PD using the Gamma knife [33]. In the first, a dose of 180 Gy with an 8-mm collimator was used. This patient had complete relief of his tremor 1 month after the lesion. Unfortunately, after 6 months a right hemiparesis and dysphasia ensued. After a course of steroids and observation the patient was left with a mild hemiparesis and no tremor. The second patient received 200 Gy with a single 4-mm collimator and obtained only a transient relief of the tremor.

Another 3 patients were treated by Friehs et al. [12] using a single 4-mm collimator with a maximum dose of 160 Gy. All 3 patients showed good response 3 or 4 weeks postradiosurgery with partial or complete resolution of their tremor; however, the follow-up was only 1 year in duration.

Hirato et al. [14] treated 1 patient using Gamma knife thalamotomy with 150 Gy maximum dose, and the patient noted improvement in her tremor 3 months after treatment, and it was markedly diminished 6 months later without complications.

Patient Selection. Of the 1,000 patients treated using Gamma knife radiosurgery at Good Samaritan Hospital over the past 6 years, Gamma thalamotomies represent only 3.4% of the total cases. They also represent only 6% of the total number of PD cases treated at our institution. Our results in treating the tremor of PD using radiofrequency lesioning with unit cell recording and physiological feedback is 90% complete tremor relief, with a less than 3% morbidity, thus, this is usually our procedure of choice for this group of patients. This retrospective review proved that for a small subset of patients not normally considered for surgical intervention, or who preferentially chose this course of treatment, Gamma knife thalamotomy has value.

Target Selection. With more and more electrophysiological knowledge of the basal ganglia, targeting has changed over the decades. At first, the efferent pathway from the globus pallidus to the ventralis anterialis oralis (VOA) of the thalamus was thought to be the prime target for tremor elimination. Since then, lesions have moved posteriorly to the VIM for selective thalamotomy for tremor [43].

Landmarks using the AC/PC line were used for targeting the contralateral VIM nucleus. In addition, subjective, visual inspection of the MRI image and Schaltenbrand atlas [42] (anatomical landmarks) were also used in the decision process for placement of the target in order to account for patient anatomical variation. Two-millimeter thick MRI slices with T2- and STIR-weighted sequences give excellent gray and white matter differentiation which we found correlated well with surgical anatomy. By visual inspection only, follow-up MRIs showed that the planned targets and the actual lesions coincided. In no patients did the lesion itself stray into the internal capsule.

Based on this series, peak lesion visualization and edema patterns became evident within 1 year of follow-up. MRI follow-ups thereafter are not necessary.

Radiologic Findings. The size of the radiosurgical lesions on gadolinium-enhanced T1-weighted sequences between the high- and low-dose groups were not statistically different and were consistent between patients. This was not the case for the T2-weighted sequences. Although not statistically different, there was a trend for the higher dose lesions to elicit a larger T2-weighted signal change or T2 'streaking'. This streaking may represent edema, radiation change, demyelinization or necrosis. It is unlikely that this represents necrosis in that the presence of the 'streaking' within the capsule or other thalamic nuclei never correlated with neurologic impairment. In addition, this streaking did not create mass effect. Only postmortem studies will elucidate the true nature of this interesting finding on follow-up scans. Clinical improvement in the higher dose group may be explained by a *physiologically* larger lesion in this group correcting any target-planning inaccuracies.

Radiosurgical Pallidotomy

This report represents the largest series of patients treated with a radiosurgical lesion to the region of the globus pallidus. The paucity of reports in the literature no doubt reflects a deserved lack of faith in the procedure.

The experience of Friedman et al. [11] is similar to ours. They described their results in 4 patients using Gamma knife pallidotomy in advanced disease. The selected target was that of the internal globus pallidus (GPi) as described by Laitinen and co-workers [25, 26]. All 4 patients exhibited a response to levodopa prior to inclusion in the treatment. A single 4-mm collimator with a maximum dose prescription of 180 Gy was used to make the lesion in all patients. No patient improved in a significant manner within the follow-up interval of 18 months. One patient experienced an improvement in his dyskinesia, but also became transiently psychotic and demented. The other 3 patients suffered no adverse effects. Follow-up MRI scans at 1 year revealed accurately placed lesions, but with variable and unpredicted sizes.

Target Selection. With more and more electrophysiological knowledge of the basal ganglia, targeting has changed over the decades. Early pallidotomies were performed in the anterodorsal pallidum [6, 13, 38]. Later studies have revealed that lesions of the posteroventral pallidum yielded good control of bradykinesia and tremor [25]. The posteroventral lesion location has since been targeted in a large number of PD patients and has consistently demonstrated very good results with rigidity and bradykinesia, as well as control of tremor in many reports [4, 15, 16, 18, 25, 27, 34, 35, 44]. Even within the posteroventral pallidum, however, subtle differences in lesion targeting have the potential to affect outcome. Without physiologic feedback, differentiation of internal and external globus pallidus was impossible during gamma pallidotomy, and the advantages gained by electrophysiologic unit-cell recording during radiofrequency lesioning was lost. The lack of clinical improvement may therefore have been attributable to 'sloppy' or inaccurate physiological lesioning within the globus pallidus.

Radiologic Findings. The explanation for the profoundly high complication rate of 50% in this series is no doubt due to the variability and unpredictability of the lesion size when the globus pallidus serves as the target. This unpredictability and variability was not seen in the VIM thalamotomy series [8] and probably represents anatomical susceptibility to very small vessel venous or arterial infarction in the area of GPi.

Differential Sensitivity of the Globus Pallidus to Single Fraction Radiosurgery. Bilateral pallidal lesions from carbon monoxide (CO) poisoning are well understood. The affinity of CO for the hemoglobin molecule impairs oxygen transport and release. Oxygen and CO are competitors for the same binding sites on the hemoglobin molecule. Therefore, the pallidum is sensitive to relative hypoxia. Other areas of the brain such as the hippocampus and cerebellum are also sensitive to ischemia and hypoxia, but the pallidum seems particularly so.

This special sensitivity of the pallidum has also been borne out in case reports of hypoxic injury from inhalation pneumopathy, cardiac arrest, and sepsis [10] where patients have exhibited movement disorders upon resuscitation and bilateral pallidal lesions on MRI.

'Postcardiac surgery choreic syndrome' occurs usually in children as a complication of cardiac surgery requiring deep hypothermia and bypass. Clinically the children have choreoathetosis and on MRI and CT scans they exhibit bilateral selective globus pallidus injuries [24].

Hallervorden-Spatz disease, a rare fatal disorder, characterized by choreoathetoid movements and 'tiger's eye' appearance on MRI due to pallidal necrosis, is yet another disorder suggestive of the susceptibility of the pallidum to metabolic disease [36].

Finally, the pallidum is known to contain high levels of iron, and these levels typically rise with age. It has been hypothesized that the presence of iron within this structure may catalyze free radical reactions causing toxicity to the aging brain [3].

The above evidence for special sensitivity of the pallidum to hypoxia and perhaps free radical formation speaks to the possible explanation for the high complication rate of injury to the pallidum with single-fraction radiosurgery. The tapering end-artery distribution of the lenticulostriate supply may be more susceptible to radiation vascular effects which in turn could cause infarct and edema seen in this series. With a maximum dose of 160 Gy, the 10% isodose line (16 Gy) of a 4-mm collimator may measure 12 mm in diameter. It is entirely possible that radionecrosis or tissue hypoxia or vascular damage may occur within that isodose volume at that dose.

This would also explain why certain patients had no adverse effects related to the dose. Perhaps less iron composition in their pallidum (causing less free radical formation), or better blood supply to the area, counteracted the effects of the single-fraction radiosurgery.

Summary

Gamma knife radiosurgical thalamotomy is an effective and useful alternative to invasive radiofrequency techniques for patients at high surgical risk. Higher radiosurgical doses are more effective than lower ones at eliminating or reducing tremor, and are without complications. The mechanical accuracy of the gamma unit combined with the anatomical accuracy of high-resolution MRI make radiosurgical lesioning in the thalamus safe and precise. On the other hand, the results from radiosurgical lesioning of the GPi have been disappointing. A 50% complication rate (homonymous field cuts, hemipareses and dysphagias) combined with a poor success rate (only 33% of patients showed any improvement in dyskinesia and rigidity) has led us to re-evaluate the indications for this procedure in the face of the excellent results from radiofrequency pallidotomy with physiological monitoring. Age-related or anatomy-related susceptible blood supply to the area may lead to hypoxia after single-fraction radiosurgery, in a nuclear complex known to be specially susceptible to hypoxia. In addition, varying levels of iron deposition within the pallidum may catalyze free radical formation in the elderly only to be further exacerbated by tissue hypoxia. At our institution, we have a moratorium on the performance of radiosurgical pallidotomies. Perhaps much lower doses will be capable of producing the wanted effects of pallidotomy in the future.

References

1   Bakay RAE, DeLong MR, Vitek JL: Posteroventral pallidotomy for PD. J Neurosurg 1992;77: 487–488.
2   Barcia Salorio JL, Roldan P, Hernandez G, et al: Radiosurgical treatment of epilepsy. Appl Neurophysiol 1985;48:400–403.
3   Bartozokis G, Beckson M, Hance D, et al: MR evaluation of age-related increase of brain iron in young adult and older normal males. Magn Reson Imaging 1997;15:29–35.
4   Ceballos-Baumann AO, Obeso JA, Vitek JL, Delong MR, Bakay R, Linazasoro G, Brooks DJ: Restoration of thalamocortical activity after posteroventral pallidotomy in PD. Lancet 1994;344: 814.
5   Cooper IS, Bravo C-J: Implications of a five-year study of 700 basal ganglia operations. Neurology 1958;8:701–707.
6   Cooper IS, Bravo G: Chemopallidectomy and chemothalamectomy. J Neurosurg 1958;15:244–250.
7   Dahlin H, Larsson B, Leksell L, et al: Influence of absorbed dose and field size on the geometry of the radiation-surgical brain lesion. Acta Radiol Ther Phys Biol 1975;4:139–145.
8   Duma CM, Jacques DB, Mark R, et al: Gamma Knife radiosurgical thalamotomy for Parkinson's tremor: A five-year experience. Neurosurg Focus 1997;(3):Article 12.
9   Fahn S, Elton RL, Members of the UPDRS Development Committee. Unified Parkinson's disease ration scale; in Fahn S, Marsden CD, Caine DB, Goldstein M (eds): Recent Developments in Parkinson's Disease. Florham Park/NJ, MacMillan Healthcare Information, 1987, vol 2, pp 153–163.
10 Feve A, Fenelon G, Wallays C, et al: Axial motor disturbances after hypoxic lesions of the globus pallidus. Mov Disord 1993;8:321–326.
11 Friedman JH, Epstein M, Sanes JN, et al: Gamma Knife pallidotomy in advanced PD. Ann Neurol 1996;39:535–538.
12 Friehs GM, Ojakangas CL, Pachatz P, et al: Thalamotomy and caudatotomy with the gamma knife as a treatment for parkinsonism with a comment on lesion sizes. Stereotact Funct Neurosurg 1995; 64:209–221.
13 Guiot G, Brion S: Traitement des mouvements anormaux par la coagulation pallidale. Technique et résultats. Rev Neurol 1953;89:578–580.
14 Hirato M, Ohye C, Shibazaki T, et al: Gamma Knife thalamotomy for the treatment of functional disorders. Stereotact Funct Neurosurg 1995;64:164–171.
15 Iacono RP, Lonser RR, Oh A, Yamada S: New pathophysiology of PD revealed by posteroventral pallidotomy. Neurol Res 1995;17:178–180.
16 Iacono RP, Shima F, Lonser RR, Kuniyoshi S, Maeda G, Yamada S: The results, indications, and physiology of posteroventral pallidotomy for patients with PD. Neurosurgery 1995;36:118–127.
17 Iacono RP, Shima F, Louser R, et al: Results and mechanisms for pallidotomy for Parkinson's akinesia. Mov Disord 1994;9:484–485.
18 Jankovic J, Cardoso F, Grossman RG, et al: Outcome after stereotactic thalamotomy for parkinsonian, essential and other types of tremor. Neurosurgery 1995;17:680–687.
19 Kelly PJ, Ahlskog JE, Goerss SJ, et al: Computer-assisted stereotactic ventralis lateralis thalamotomy with microelectrode recording control in patients with Parkinson's disease. Mayo Clin Proc 1987; 62:655–664.
20 Kondziolka D, Linskey ME, Lunsford LD: Animal models in radiosurgery; in Alexander E III, Loeffler J, Lunsford LD (eds): Stereotactic Radiosurgery. New York, McGraw-Hill, 1993, pp 51–64.
21 Kondziolka D, Lunsford LD, Claasen D, et al: Radiobiology of radiosurgery. I. The normal rat brain model. Neurosurgery 1992;31:940–945.
22 Kondziolka D, Lunsford LD, Flickinger JC, et al: Stereotactic radiosurgery for trigeminal neuralgia: A multi-institutional study using the gamma unit. J Neurosurg 1996;84:940–945.
23 Kopyov O, Jacques DS, Duma CM, Buckwalter G, Kopyov A, Lieberman A, Copcutt BG: Microelectrode-guided posteroventral radiofrequency pallidotomy for PD provides symptom relief with few adverse effects. J Neurosurg, 1997;87:52–59.
24 Kupsky W, Drozd M, Barlow C: Selective injury of the globus pallidus in children with post-cardiac surgery choreic syndrome. Dev Med Child Neurol 1995;37:135–144.

25 Laitinen L, Bergenheim AT, Hariz MI: Leksell's posteroventral pallidotomy in the treatment of PD. J Neurosurg 1992;76:53–61.

26 Laitinen L: Brain targets in surgery for Parkinson's disease. J Neurosurg 1985;62:349–351.

27 Lehman RM, Mesrich R, Sage J, Goldbe L: Pre- and postoperative magnetic resonance imaging localization of pallidotomy. Stereotact Funct Neurosurg 1994;62:61–70.

28 Leksell L, Backlund EO: Stereotactic gammacapsulotomy; in Hitchcock ER, Ballantine HT Jr, Meyerson BA (eds): Modern Concepts in Psychiatric Surgery. Amsterdam, Elsevier, 1979, pp 213–216.

29 Leksell L: Cerebral radiosurgery. I. Gamma thalamotomy in two cases of intractable pain. Acta Chir Scand 1968;134:585–595.

30 Leksell L: Stereotactic radiosurgery in TN. Acta Chir Scand 1971;137:311–314.

31 Leksell L: The stereotaxic method and radiosurgery of the brain. Acta Chir Scand 1951;102:316–319.

32 Lindquist C, Kihlstrom L, Hellstrand E: Functional neurosurgery – A future for the gamma knife? Stereotact Funct Neurosurg 1991;57:72–81.

33 Lindquist C, Steiner L, Hindmarsh T: Gamma Knife thalamotomy for tremor: Report of two cases; in Steiner L (ed): Radiosurgery, Baseline and Trends. New York, Raven Press, 1992, pp 37–243.

34 Lozano A, Hutchison W, Kiss Z, Tasker R, Davis K, Dostrovsky J: Methods for microelectrode-guided posteroventral pallidotomy. J Neurosurg 1996;84:194–202.

35 Lozano A, Lang A, Galvez-Jiminez N, et al: GPi pallidotomy improves motor function in patients with PD. Lancet 1995;346:1383–1386.

36 Malandrini A, Fabrizi G, Bartalucci P, et al: Clinicopathological study of familial late infantile Hallervorden-Spatz disease: A particular form of neuroacanthocytosis. Childs Nerv Syst 1996;12:155–160.

37 Meyers HR: Surgical procedure for postencephalitic tremor with notes on the physiology of premotor fibers. Arch Neurol Psychiatry 1940;44:453–459.

38 Narabayashi H, Okuma T: Procaine-oil blocking of the globus pallidus for the treatment of rigidity and tremor of parkinsonism. Proc Jpn Acad 1953;29:134–137.

39 Ohye C: Selective thalamotomy for movement disorders: Microrecording stimulation techniques and results; in Lunsford LD (ed): Modern Stereotactic Neurosurgery. Boston, Nijhoff, 1988, pp 315–331.

40 Otsuki T, Jokura H, Takahashi K, et al: Stereotactic gamma-thalamotomy with a computerized brain atlas: Technical case report. Neurosurgery 1994;35:764–767.

41 Rand RW, Jacques DB, Melbye RW, Copcutt BG, Levenick MN, Fisher M: Gamma knife thalamotomy and pallidotomy in patients with movement disorders: Preliminary results. Stereotact Funct Neurosurg 1993;61(suppl 1):65–92.

42 Schaltenbrand G, Wahren W: Atlas for Stereotaxy of the Human Brain, ed 2. Chicago, Year Book Medical. Stuttgart, Thieme, 1977, plates 16–18.

43 Steiner L, Forster D, Leksell L, et al: Gamma thalamotomy in intractable pain. Acta Neurochir 1980;52:173–184.

44 Sutton JP, Couldwell W, Lew MF, Mallory L, Grafion S, DeGiorgio C, Welsh M, Apuzzo ML, Ahmadi J, Waters CH: Ventroposterior medial pallidotomy in patients with advanced PD. Neurosurgery 1995;36:1112–1167.

45 Tasker RR: Thalamotomy. Neurosurg Clin North Am 1990;1:841–864.

Christopher M. Duma, MD, Neurosciences Institute, Good Samaritan Hospital,
637 South Lucas Avenue, Suite 501, Los Angeles, CA 90017 (USA)
Tel. (213) 977 2234, Fax (213) 482 2157

Lunsford LD, Kondziolka D, Flickinger JC (eds): Gamma Knife Brain Surgery.
Prog Neurol Surg. Basel, Karger, 1998, vol 14, pp 212–221

....

Gamma Knife Radiosurgery for Trigeminal Neuralgia

Douglas Kondziolka, Bernardo Perez, John C. Flickinger, L. Dade Lunsford

Departments of Neurological Surgery and Radiation Oncology and
The Center for Image-Guided Neurosurgery, University of Pittsburgh
Medical Center, Pittsburgh, Pa., USA

Medical management remains the primary treatment for trigeminal neuralgia [1]. Several effective surgical techniques can be offered to patients when
medical therapy is ineffective or associated with significant side effects [2–10].
Although most procedures have high rates of initial success, most physicians
and patients choose medical therapy first because of potential surgical morbidity, the risk for loss of facial sensation after surgery, or recurrent pain despite
initial surgical success. These procedures include percutaneous radiofrequency
rhizotomy, glycerol rhizotomy, mechanical balloon compression, peripheral
nerve section, and microvascular decompression. Despite these options, significant problems continue to challenge the management of this pain syndrome.
These problems range from persistent or recurrent typical trigeminal neuralgia
despite medical or surgical therapy, to the tendency for pain to occur in elderly
patients. Elderly patients are prone to have other medical illnesses that warrant
a minimally invasive approach.

Initially, our goal was to re-evaluate the early anecdotal success of trigeminal neuralgia radiosurgery reported by Leksell [11]. During the 1970s and
1980s, several surgeons irradiated the trigeminal ganglion [12–14] using the
gamma knife. In 1996 we reported a multicenter study using Gamma knife
radiosurgery for radiation of the proximal trigeminal nerve near the pons [15].
At this location, the nerve could be imaged with high-resolution magnetic
resonance imaging (MRI) and was suitable for small-volume radiosurgery
targeting. In that 50-patient study with an 18-month median follow-up, 58%
were pain-free, 36% had significant improvement (50–90% relief) and 6% failed.
The main purpose of that study was to identify an appropriate radiation dose

which might eliminate pain more rapidly and still maintain facial sensation. We found that a maximum radiosurgery dose of >70 Gy (70–90 Gy) was associated with a greater chance of complete pain relief than 60 or 65 Gy ($p = 0.0003$)[15]. Herein we report our longer-term experience in patients with typical trigeminal neuralgia who had Gamma knife radiosurgery and evaluate a large patient series who received a maximum dose of 70–90 Gy with longer-term follow-up.

University of Pittsburgh Series

One hundred and seventy-two patients underwent Gamma knife radiosurgery at our center between 1993 and 1997. The mean patient age was 67 years (range 21–92). No patient had an associated mass lesion. The dose range used was 60–90 Gy, initially beginning with a dose-escalation protocol.

We evaluated 121 patients with typical trigeminal neuralgia (age range 32–92). All patients underwent comprehensive trials of medical therapy which included carbamazepine. Many patients also received phenytoin, baclofen, and/or gabapentin, alone or in combination with carbamazepine. Prior surgeries included microvascular decompression ($n = 42$), glycerol rhizotomy ($n = 57$) or radiofrequency rhizotomy ($n = 19$). No patient had anesthesia dolorosa. The right side of the face was involved in 72 patients (59.5%) and the left side in 49. Pain involved the V_1, V_2 or V3 distributions, alone or in combination. The most common distribution was combined V_2 and V_3 pain. Forty-six patients described some degree of facial numbness.

Radiosurgery Technique

All patients underwent stereotactic radiosurgery using the Leksell Gamma knife (Elekta Instruments, Atlanta, Ga., USA). The Model U Gamma knife was used in 79 patients and the Model B in 42 patients. The dose profile of the 4-mm isocenter is only slightly different between the two units with the superior-inferior radiation volume being slightly larger in the U unit. Radiosurgery was performed under local anesthesia.

After application of the Leksell Model G stereotactic frame (Elekta Instruments), all patients underwent stereotactic MRI to identify the trigeminal nerve. MRI sequences were performed using short repetition time (TR) sequences and contrast enhancement in multiple planes. Volume acquisition sequences using 512×256 matrices were divided into 1-mm image slices to provide graphic depiction of the trigeminal nerve (fig. 1). The nerve was identified as it followed its course from the brainstem into Meckel's cave. In some patients who had undergone prior surgery (microvascular decompression

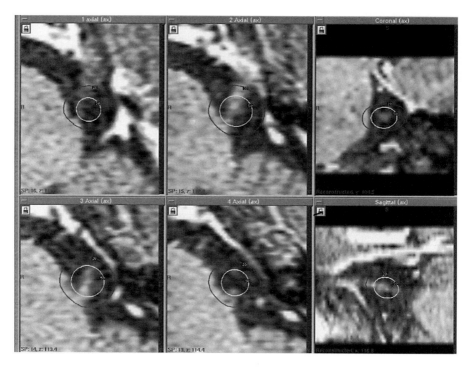

Fig. 1. Dose-planning MR images in an 80-year-old woman with left trigeminal neuralgia. A single 4-mm isocenter was targeted to the proximal trigeminal nerve to deliver a maximum dose of 75 Gy. Coronal and sagittal views are shown on the right.

or percutaneous surgery) the nerve was difficult to identify (either because of nerve atrophy or regional perineural fibrosis) and thus long TR MR sequences were used to identify the nerve against the high-signal background of cerebrospinal fluid. The 4-mm isocenter was placed 2–4 mm anterior to the junction of the trigeminal nerve and pons (fig. 2). The isocenter was positioned so that the brainstem surface was usually irradiated at no more than the 30% isodose. The range of maximum radiosurgery dose was between 70 and 90 Gy. 70 Gy commonly was used for patients who had not undergone prior surgery and 80–90 Gy was prescribed for patients with recurrent pain after prior surgery, although a strict protocol did not exist.

All patients were discharged within 24 h after radiosurgery. Patients were studied according to the degree of pain relief, latency interval to pain relief, onset of paraesthesia, need for further surgical treatment, and complications. Patient evaluations were made by a physician who did not participate in the procedure, and who was blinded to radiation dose.

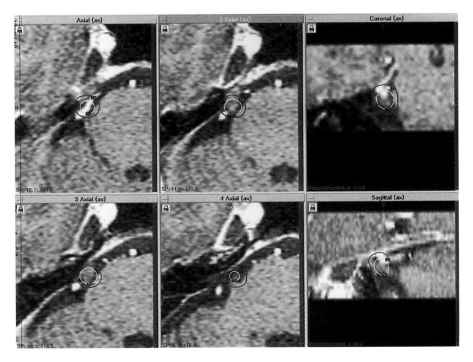

Fig. 2. Radiosurgery images in a 68-year-old man with right trigeminal neuralgia. The 4 mm isocenter was placed so that the 20% isodose line was just at the brainstem surface. A maximum dose of 75 Gy was delivered. Note the vascular compression superior to the nerve.

Trigeminal Neuralgia Pain Relief

Median follow-up after radiosurgery was 18 months (range 6–48). Relief from pain was coded by the patients into three categories [12, 13, 30]. These included no response, slight improvement (10–50% improved), good response (50–90% improved but still using medications if used preoperatively), and excellent response (100% pain-free, off medication). No response or slight improvement was referred to as a treatment failure. Criteria for improvement included a reduction in both the frequency and severity of trigeminal neuralgia attacks. Patients with good results continued to take medication therapy (although usually reduced) if they had done so before radiosurgery. The median time to response was 4 weeks (range 1 day to 3 months). Of 121 patients, 106 were evaluable in that complete follow-up pain relief and morbidity data was obtained.

Initial improvement in trigeminal neuralgia was noted in 91 patients (86%). These included 64 patients who had an excellent response, 18 who had a good response, 9 who were slightly improved, and 15 with no improvement. At last follow-up, significant pain relief was noted by 77% of patients (good plus excellent results). Relapse in pain was noted in only 6 of 64 patients who attained complete relief (10%) (occurred within 10 months after onset of complete relief).

Pain Response and Radiosurgery Dose

A maximum dose of 70 Gy was delivered to 54 patients. Five patients had multiple sclerosis. Sixty patients received a maximum dose of 80 Gy and 5 of these patients had multiple sclerosis. Five patients received a maximum dose of 85 Gy and 2 a maximum dose of 90 Gy. One of these patients had multiple sclerosis and the other multiple prior surgeries. There was no significant difference in pain relief when we compared 70 Gy to >80 Gy ($p = 0.57$ for complete pain relief and $p = 0.37$ for good plus excellent result).

Patient Factors and Pain Relief

We compared specific patient factors with degree of pain relief. For onset of complete pain relief, age proved insignificant ($p = 0.75$) as did sex ($p = 0.22$), or history of prior surgery ($p = 0.34$). These factors also were insignificant for the onset of any improvement in pain. A history of facial numbness before radiosurgery did not correlate with pain relief ($p = 0.32$). We also compared degree of pain relief with the Gamma knife model used. No difference was identified ($p = 0.46$). The Model B unit has a greater radiation volume in the left-right dimension and the Model U unit, in the superior-inferior dimension. We also coded the surgeons' ability to identify clearly the nerve on high-resolution MR images. Poorer trigeminal nerve resolution was found in some patients ($n = 11$) with multiple prior surgeries. However, better nerve identification did not correlate with pain relief ($p = 0.62$). None of these factors were significant in multivariate analysis.

Facial Paresthesiae

Twelve patients developed new or increased trigeminal paresthesiae or sensory loss (numbness) after radiosurgery (10%). Because many patients lived at a distance from Pittsburgh, we were not able to perform a detailed sensory examination and code this finding but relied on a patient survey. The factors of age, multiple sclerosis, sex, radiosurgery dose, Gamma knife model, or nerve identification did not correlate with the onset of sensory findings ($p > 0.05$). There was a trend for a lower rate of sensory findings in patients with no prior surgery ($p = 0.08$). No patient developed anesthesia dolorosa after radiosurgery.

Other Morbidity

All patients were discharged from hospital within 24 h. Rare patients had an increase in trigeminal neuralgia pain within the first few hours after radiosurgery that may have been related to an acute radiation effect. No patient developed nausea or headache after stereotactic radiosurgery. No patient developed new neurologic deficits or other systemic complications.

History of Trigeminal Neuralgia Radiosurgery

Stereotactic radiosurgery was conceived for the management of functional brain disorders. Leksell and Larsson developed the Gamma knife as a method for cerebral lesion generation without opening the skin and skull. Initial functional applications included lesion generation for Parkinson's disease (ventrolateral thalamotomy), pain (medial thalamotomy), affective disorders (anterior capsulotomy and cingulotomy), trigeminal neuralgia, and later epilepsy [12, 13, 18–20]. As a small-volume lesion generator, the Gamma knife has proved to be a reliable tool. Single-session doses from 50 to 200 Gy can cause tissue necrosis within 6 months, depending on target volume [21, 22]. Smaller volumes require more radiation. There exists a therapeutic range where nerve or brain parenchyma can be affected without requiring total parenchymal injury. This range appears to differ with brain location, target volume and desired response time.

Radiosurgery is the least invasive surgical procedure for trigeminal neuralgia [15, 23]. We and others have noted no systemic morbidity. No patient sustained any form of neurologic morbidity other than a low risk for facial numbness. In the multi-institution study, 3 of 50 patients (6%) developed increased facial paresthesia after radiosurgery. Young et al. [23] noted that only 1 of their 60 patients had increased facial sensory loss and this occurred in a patient with a trigeminal schwannoma. In the present study we found a 10% rate of partial facial numbness. We initially believed this slightly higher rate was due to use of some higher radiosurgery doses (80–90 Gy) and selection of patients who already had some numbness, although these factors did not prove to be significant. The absence of infection, cerebrospinal fluid leakage, anesthesia complications, hearing loss, facial hematoma, facial weakness or brainstem injury has established radiosurgery as an attractive surgical alternative for many patients.

Despite these advantages, the use and evaluation of radiosurgery has proceeded slowly. Leksell [11, 24] first used radiosurgery techniques for trigeminal nerve irradiation in 1953. Results from these 2 patients (both were pain-free after delays of 1 in 5 months) were not published until 1971. Leksell [11] concluded that, 'from these observations no definite conclusion should be drawn concerning the optimal dose of radiation or the exact mechanism in site of action in the route or ganglion, or even the general applicability of the method'. In 1983, Leksell [12] noted that 63 patients had had Gamma knife radiosurgery for trigeminal neuralgia. He did not describe the surgical method or results. The use of gasserian ganglion radiosurgery was reported by Lind-quist et al. [13] (46 patients) and Rand et al. [14] (12 patients) but inconsistent results were obtained. These authors concluded that the ganglion was probably not appropriate as the primary radiosurgery target.

Gamma Knife Technique

During the last 5 years, we and others have worked to determine the appropriate patient for trigeminal neuralgia radiosurgery, the best radiation dose, and clarify expectations. The best radiation dose to relieve pain quickly, maintain that relief over the long term, and do so with no systemic morbidity and a low rate of facial sensory loss, was not known. To answer this question we formed a multicenter study group and compared data from patients who had radiosurgery at doses from 60 to 90 Gy. All groups selected the proximal nerve and root entry zone as the radiation target which could be identified on MRI and targeted with the Gamma knife. Because the nerve was myelinated by oligodendrocytes and perhaps was more sensitive to irradiation than Schwann cell myelin, we hypothesized that a stronger radiobiologic effect would occur at this portion of the nerve [15, 25, 26]. We also believed that the compact union of fibers from different nerve divisions would facilitate irradiation of a smaller volume target. We attributed the observation of good or excellent results to high-resolution identification of the trigeminal nerve near the pons and accurate radiosurgical targeting [27]. In that study, a max-imum dose of at least 70 Gy was identified for a higher rate of pain relief. Radiosurgical targeting was found to be accurate as previously identified in experimental and clinical studies [22]. From these findings, we evaluated our patient series using a standard technique and a narrow radiosurgery dose range (70–90 Gy). We currently are completing a study that compares use of one versus two radiosurgery isocenters (with two isocenters, a greater length of nerve is irradiated) to determine the effect of nerve length on pain relief (fig. 3). We also have begun a study of the histologic and ultrastructural effects of trigeminal nerve radiosurgery in a large animal model.

The Physiological Effect of Radiosurgery

It is unclear why trigeminal nerve irradiation causes relief of trigeminal neuralgia pain. We speculate that nerve irradiation leads to functional electro-physiologic block of ephaptic transmission since the majority of patients main-tain normal trigeminal function [28]. Although we do not perform follow-up imaging routinely, MR studies performed 6–24 months after radiosurgery

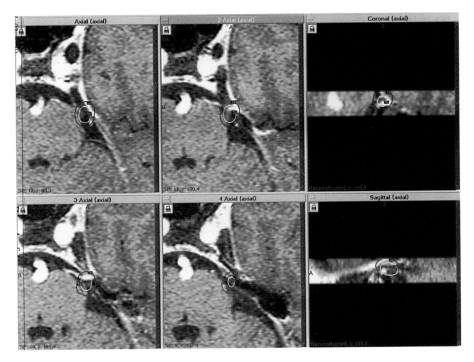

Fig. 3. Radiosurgery dose plan in a 68-year-old man who received two isocenters to create an oval radiosurgery volume for left trigeminal neuralgia. A maximum dose of 75 Gy was delivered.

show contrast enhancement at the target. Radiosurgery performed on patients with cavernous sinus tumors has been associated with low rates of trigeminal dysfunction [29]. Radiosurgery may have an effect on ephaptic transmission but not on normal axonal conduction. Young et al. [23] theorized that radiofrequency energy may interrupt both abnormal transmission and normal axonal conduction in such a way that loss of normal facial sensation is required for long-lasting pain relief. Perhaps radiation energy has a larger therapeutic ratio between pain relief and sensory loss. Recently, our group reported pain relief from sphenopalatine neuralgia after radiosurgery. In this procedure the sphenopalatine ganglion was irradiated with an 8-mm collimator to a maximum dose of 90 Gy [30].

It has been our philosophy to manage pain and yet try to maintain normal facial sensation. Dysesthetic pain syndromes and anesthesia dolorosa are disabling with few effective options. In some patients this new pain can be worse than the prior trigeminal neuralgia. For this reason, our surgical armamentarium includes procedures with lower rates of facial sensory loss (microvascular decompression, glycerol rhizotomy and stereotactic radiosurgery). We perform radiofrequency rhizotomy or nerve section only on rare patients. Radiosurgery is a minimally invasive method to manage trigeminal neuralgia that is associated with a low risk of facial paresthesiae, an approximate 80% rate of significant pain relief, and a low recurrence rate in patients who attain complete relief.

References

1   Maciewicz R, Scrivani S: Trigeminal neuralgia: Gamma radiosurgery may provide new options for treatment. Neurology 1997;48:565–566.
2   Broggi G, Franzini A, Lasio G, et al: Long-term results of percutaneous retrogasserian thermorhizotomy for 'essential' trigeminal neuralgia: Considerations in one thousand consecutive patients. Neurosurgery 1990;26:783–787.
3   Brown JA, McDaniel MD, Weaver MT: Percutaneous trigeminal nerve compression for treatment of trigeminal neuralgia: Results in 50 patients. Neurosurgery 1993;32:570–573.
4   Håkansson S: Trigeminal neuralgia treated by the injection of glycerol into the trigeminal cistern. Neurosurgery 1981;9:638–646.
5   Jannetta PJ: Trigeminal neuralgia: Treatment by microvascular decompression; in Wilkins R, Rengachary SS (eds): Neurosurgery. New York, McGraw-Hill, 1985, pp 2357–2363.
6   Kondziolka D, Lunsford LD, Bissonette DJ: Long-term results after glycerol rhizotomy for multiple sclerosis-related trigeminal neuralgia. Can J Neurol Sci 1994;21:137–140.
7   Lunsford LD: Treatment of tic douloureux by percutaneous retrogasserian glycerol injection. JAMA 1982;248:449–453.
8   Lunsford LD, Apfelbaum RI: Choice of surgical therapeutic modalities for treatment of trigeminal neuralgia: Microvascular decompression, percutaneous retrogasserian thermal, or glycerol rhizotomy. Clin Neurosurg 1985;32:319–333.
9   Sweet WH: The treatment of trigeminal neuralgia (tic douloureux). N Engl J Med 1986;315:174–177.
10 Young JN, Wilkins RH: Partial sensory trigeminal rhizotomy at the pons for trigeminal neuralgia. J Neurosurg 1993;79:680–687.
11 Leksell L: Stereotaxic radiosurgery in trigeminal neuralgia. Acta Chir Scand 1971;37:311–314.
12 Leksell L: Stereotactic radiosurgery. J Neurol Neurosurg Psychiatry 1983;46:797–803.
13 Lindquist C, Kihlstrom L, Hellstrand E: Functional neurosurgery – A future for the gamma knife? Stereotact Funct Neurosurg 1991;57:72–81.
14 Rand W, Jacques DB, Melbyer W, et al: Leksell gamma knife treatment of tic douloureux. Stereotact Funct Neurosurg 1993;61:93–102.
15 Kondziolka D, Lunsford LD, Flickinger JC, et al: Stereotactic radiosurgery for trigeminal neuralgia. A multi-institutional study using the gamma unit. J Neurosurg 1996;84:940–945.
16 Kaplan ES, Meier P: Nonparametric estimation from incomplete observation. J Am Stat Assoc 1958;53:457–480.
17 Cox DR: Regression models and life tables. J R Stat Soc Bull 1982;34:187–220.
18 Barcia Salorio JL, Roldan T, Hernandez G, et al: Radiosurgical treatment of epilepsy. Appl Neurophysiol 1985;48:400–403.
19 Leksell L, Backlund EO: Stereotactic gammacapsulotomy; in Hitchcock ER, Ballantine HT, Meyerson BA (eds): Modern Concepts in Psychiatric Surgery. Amsterdam, Elsevier, 1979, pp 213–216.
20 Steiner L, Forster D, Leksell L, et al: Gammathalamatomy in intractable pain. Acta Neurochir 1980;52:173–184.
21 Kondziolka D, Linskey ME, Lunsford LD: Animal models in radiosurgery; in Alexander E, Loeffler J, Lunsford LD (eds): Stereotactic Radiosurgery. New York, McGraw-Hill, 1993, pp 51–64.

22   Kondziolka D, Lunsford LD, Claassen D, et al: Radiobiology of radiosurgery. I. The normal rat brain model. Neurosurgery 1992;31:271–279.
23   Young RF, Vermeulen SS, Grimm P, et al: Gamma knife radiosurgey for treatment of trigeminal neuralgia. Idiopathic and tumor related. Neurology 1997;48:608–614.
24   Leksell L: The stereotaxic method and radiosurgery of the brain. Acta Chir Scand 1951;102:316–319.
25   Mastaglia FL, McDonald WI, Watson JV, et al: Effects of X-radiation on the spinal cord: An experimental study of the morphological changes in central nerve fibers. Brain 1976;99:101–122.
26   Van der Kogel AJ: Central nervous system radiation injury in small animal models; in Gutin P, Leibel S, Sheline G (eds): Radiation Injury to the Nervous System. New York, Raven Press, 1991, pp 91–111.
27   Kondziolka D, Dempsey PK, Lunsford LD, et al: A comparison between magnetic resonance imaging and computed tomography for stereotactic coordinate determination. Neurosurgery 1992; 30:402–407.
28   Kondziolka D, Lunsford LD, Habeck M, et al: Gamma knife radiosurgery for trigeminal neuralgia. Neurosurg Clin North Am 1997;8:79–85.
29   Mehta MP, Kinsella T: Cavernous sinus cranial neuropathies: Is there a dose-response relationship following radiosurgery? Int J Radiat Oncol Biol Phys 1993;27:477–480.
30   Pollock B, Kondziolka D: Stereotactic radiosurgery treatment of sphenopalatine neuralgia. J Neurosurg 1997;87:450–453.

Douglas Kondziolka, MD, University of Pittsburgh Medical Center, Suite B-400,
Department of Neurological Surgery, 200 Lothrop Street, Pittsburgh, PA 15213 (USA)
Tel. (412) 647 6782, Fax (412) 647 0989

Author Index

Subject Index